More Geometry Snacks

Ed Southall and Vincent Pantaloni

About the Authors

Ed Southall is a teacher trainer at Huddersfield University in the UK, and the author of *Yes But Why? Teaching for Understanding in Mathematics*. Ed has taught mathematics for 14 years in secondary schools both in the UK and the Middle East. He has developed a large following online as @solvemymaths on twitter where he regularly posts mathematical puzzles.

Vincent Pantaloni has been a French mathematics high school teacher for 18 years. After being a teacher trainer and member of research groups, he is now a mathematics inspector. He believes that searching and sharing strategies for complex problems in teamwork helps students in building a stronger mathematical knowledge. He posts about mathematics on twitter @panlepan and in French on his website Mathzani.

ISBN (book): 978-1-911093-96-1
ISBN (ebook): 978-1-911093-97-8

Printed and designed in the UK

Published by Tarquin
Suite 74, 17 Holywell Hill
St Albans AL1 1DT
United Kingdom

info@tarquingroup.com
www.tarquingroup.com

Introduction

More Geometry Snacks is the follow up volume to the hugely successful Geometry Snacks. Both are mathematical puzzle books filled with geometrical figures and questions designed to challenge, confuse and ultimately enlighten enthusiasts of all ages. Each puzzle is carefully designed to draw out interesting phenomena and relationships between the areas and dimensions of various shapes. Furthermore, unlike most puzzle books, the authors offer multiple approaches to solutions so that once a puzzle is solved, there are further surprises, insights and challenges to be had. As a teaching tool, Geometry Snacks enables teachers to promote deep thinking and debate over how to solve geometry puzzles. Each figure is simple, but often deceptively tricky to solve – allowing for great classroom discussions about ways in which to approach them.

By offering numerous solution approaches, the book also acts as a tool to help encourage creativity and develop a variety of strategies to chip away at problems that often seem to have no obvious way in.

Contents

Publisher's Note

Geometry Snacks and the Tarquin eReader

The first *Geometry Snacks* was a wonderful addition to Tarquin's range of puzzle, activity and enriching books for lovers of mathematics. This second volume adds more of the same – and also puzzles in new areas.

Both books are available in traditional book form and as ebooks though the usual outlets. We are also delighted to provide a version of the book in the Tarquin PDF Reader which is designed for teachers and those curious to explore the Snacks and their solutions using demonstrations.

Perfect for classroom use on whiteboard or computer, as well possibly offering additional insights into the problems through collaborative discussion.

For more about this, or to explore our large range of books, posters and resources for all ages go to www.tarquingroup.com or contact us at info@tarquingroup.com or @tarquingroup on Twitter.

What Fraction is Shaded?

Dedicated to Alex Bogomolny
(1964–2018)

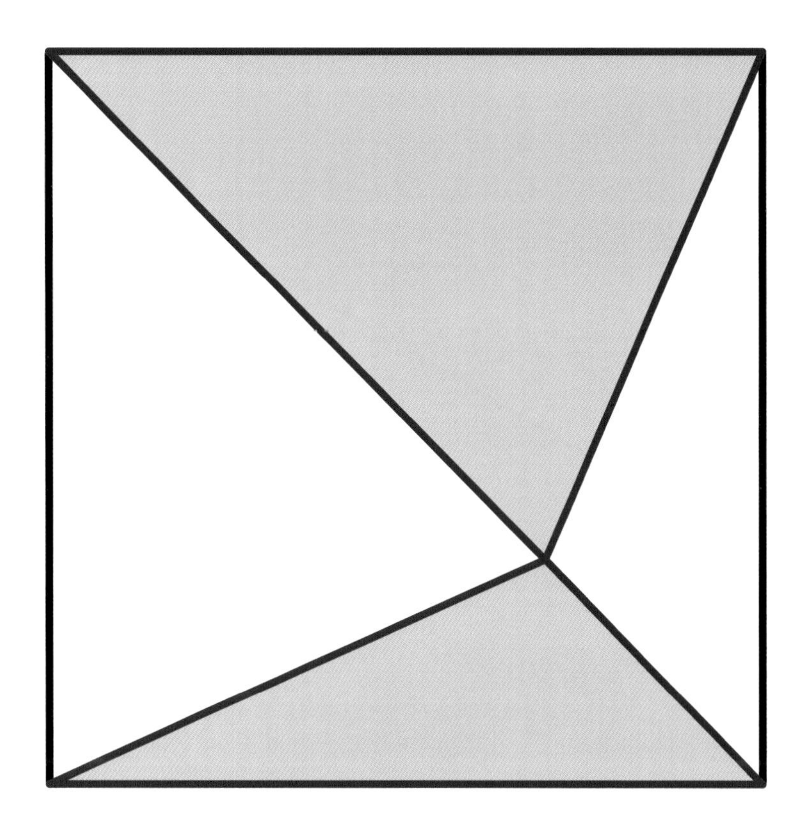

A point inside a square is connected to its four vertices.

What fraction of the square is shaded?

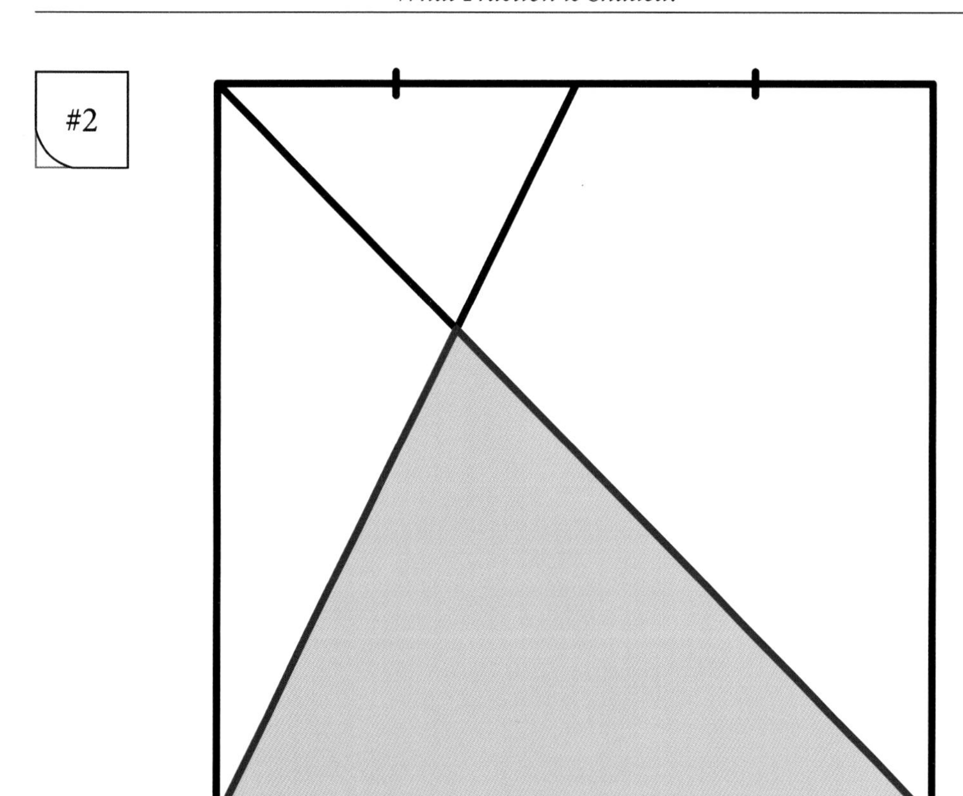

A square is segmented using the midpoint of a side, and three vertices as shown.

What fraction is shaded?

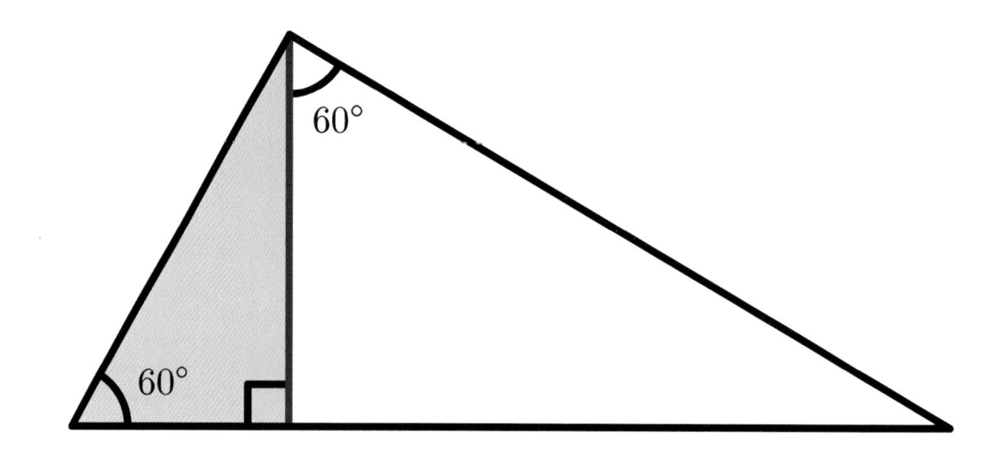

A triangle is constructed as shown.

What fraction is shaded?

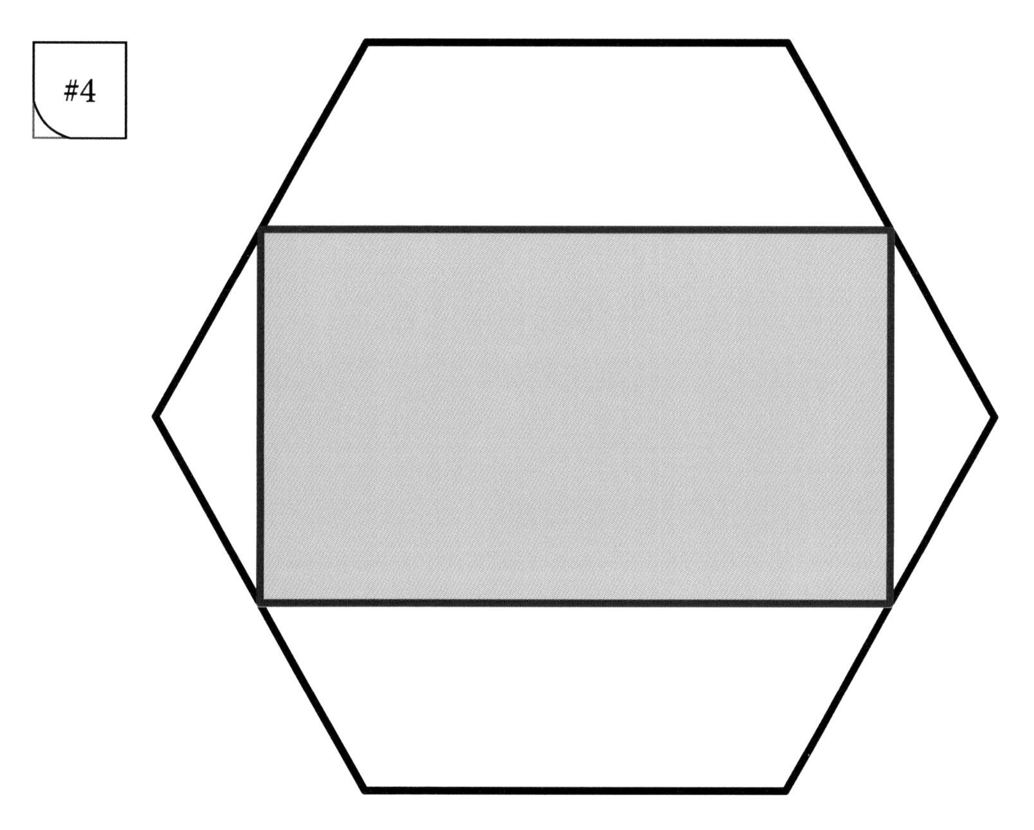

A rectangle is constructed from the midpoints of the sides of a regular hexagon as shown.

What fraction is shaded?

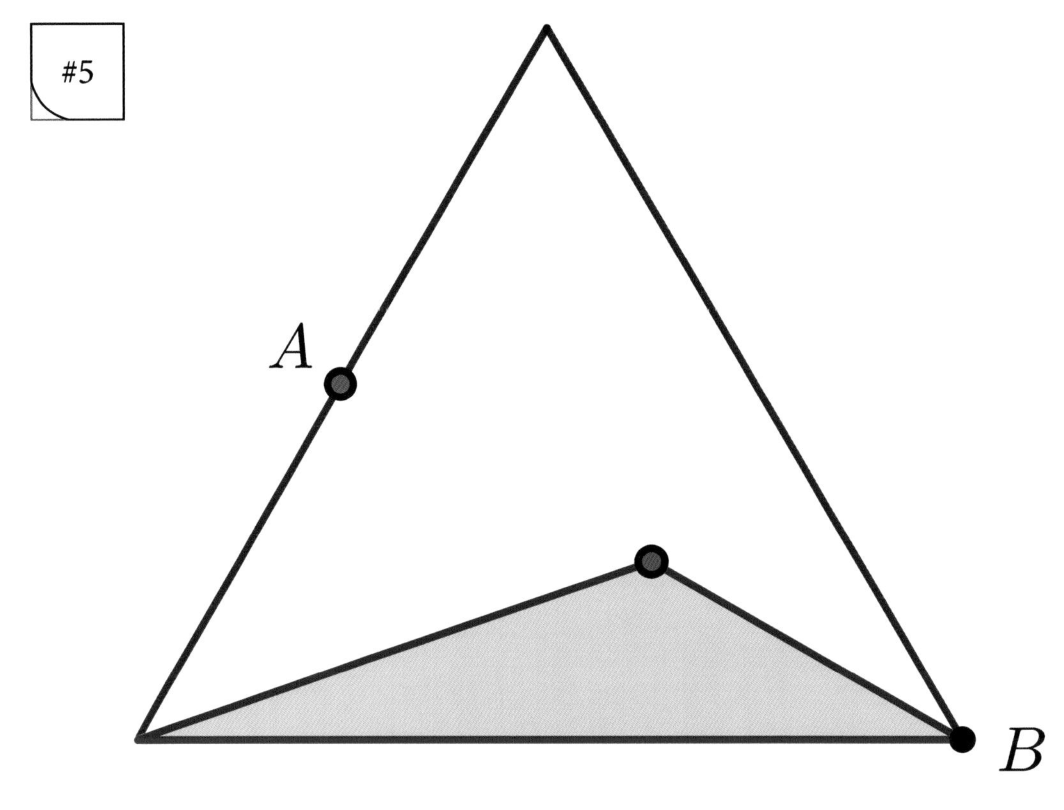

A shaded region is constructed within an equilateral triangle using the midpoint A of a side, and the midpoint between AB.

What fraction is shaded?

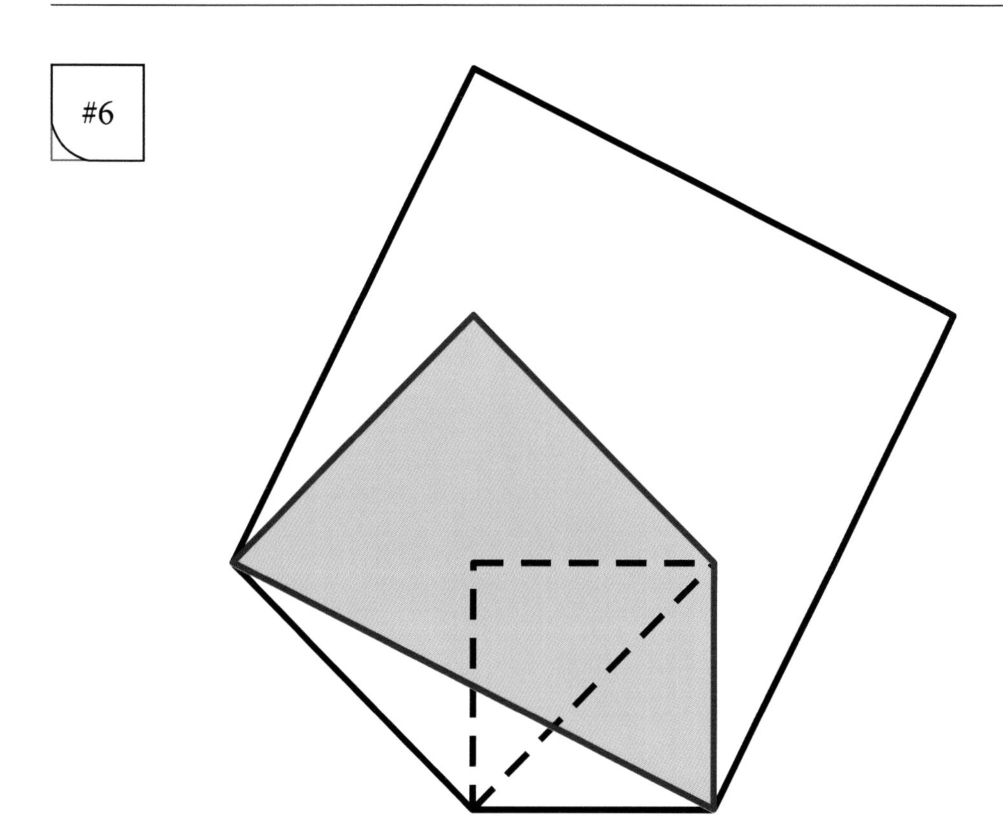

A square is constructed from two smaller squares as shown.

What fraction of the largest square is shaded?

A six petaled rosette is constructed using unit circles, and is enclosed by a rectangle as shown.

What fraction of the rectangle is shaded?

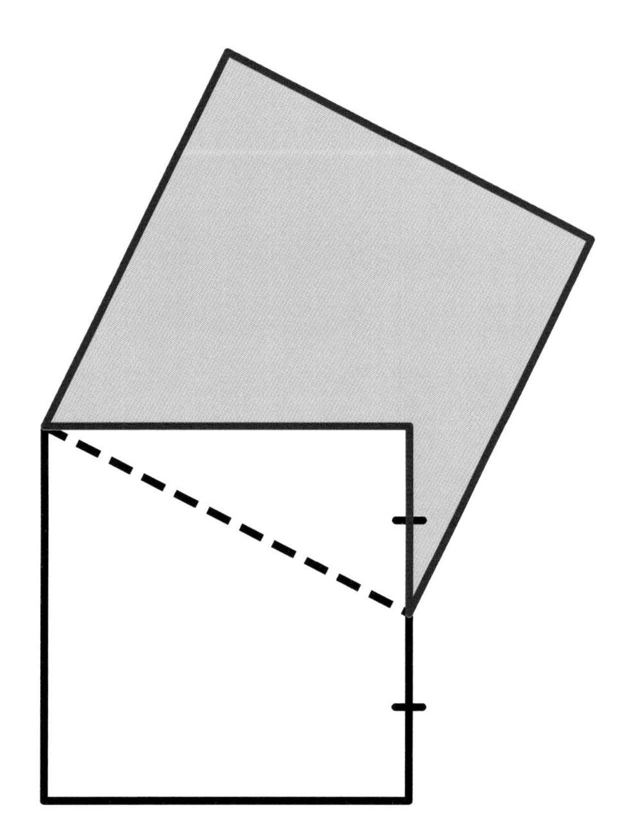

A square is formed using the vertex and midpoint of a side of a smaller square as shown.

What fraction of the irregular hexagon formed by the overlapping squares is shaded?

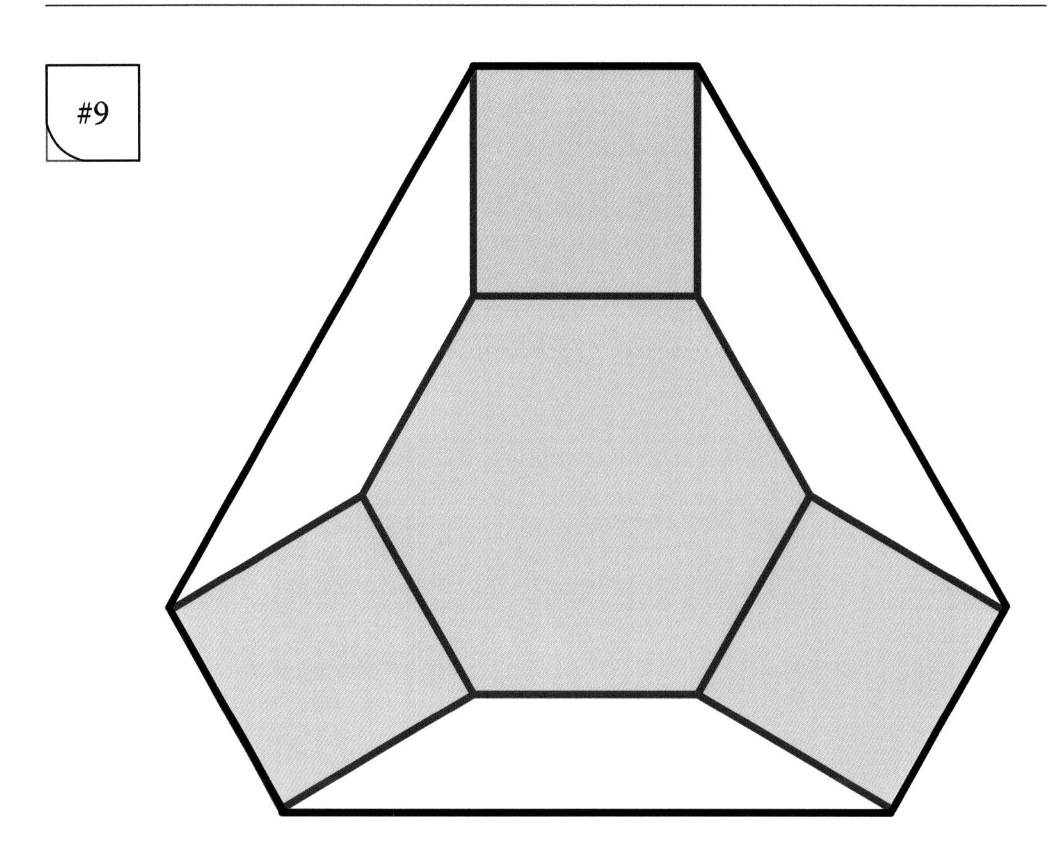

A regular hexagon with three squares on its sides is inscribed in an irregular hexagon by connecting vertices as shown.

What fraction of the figure is shaded?

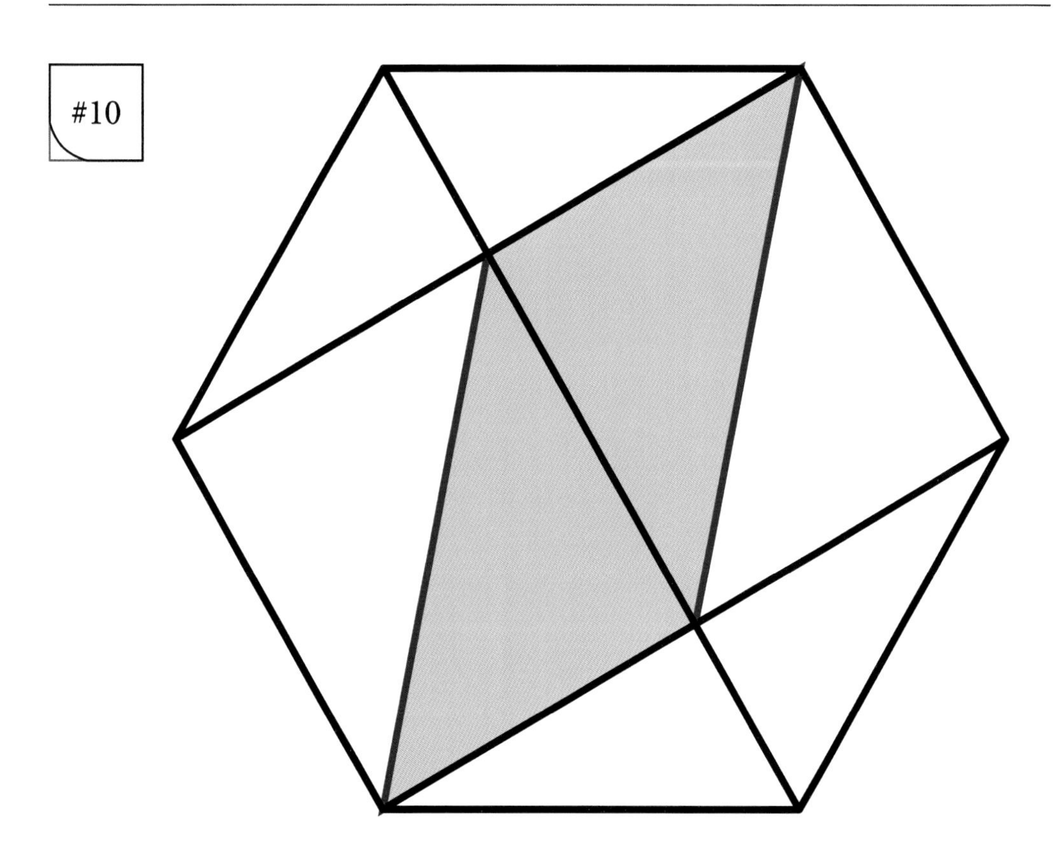

In this regular hexagon, three diagonals have been drawn to form this parallelogram.

What fraction of the hexagon is shaded?

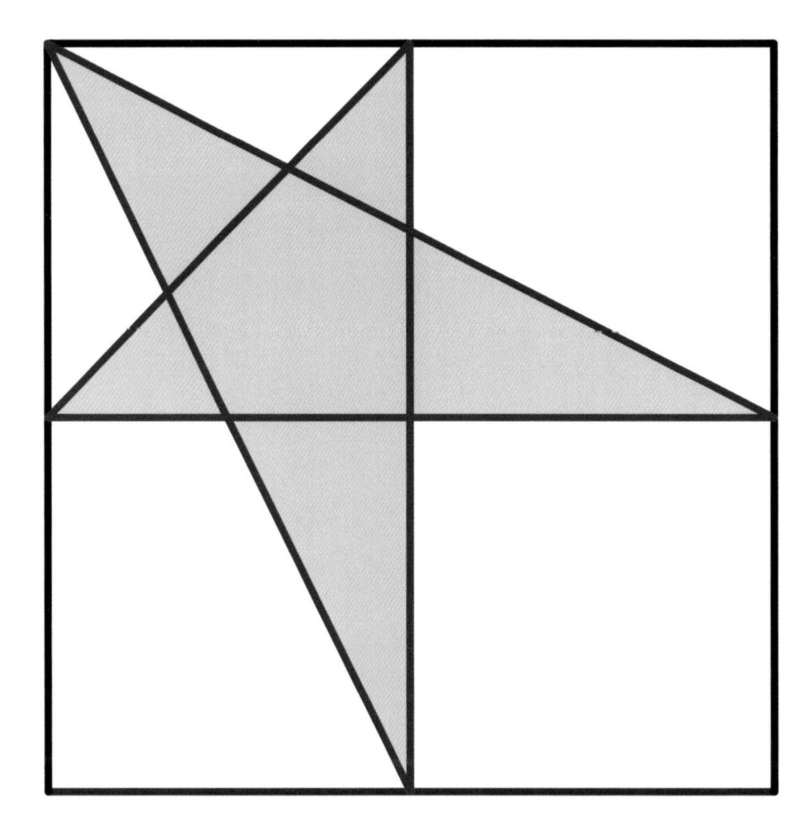

A five-pointed star is formed in a square by joining a vertex and its four midpoints as shown.

What fraction of the square is shaded?

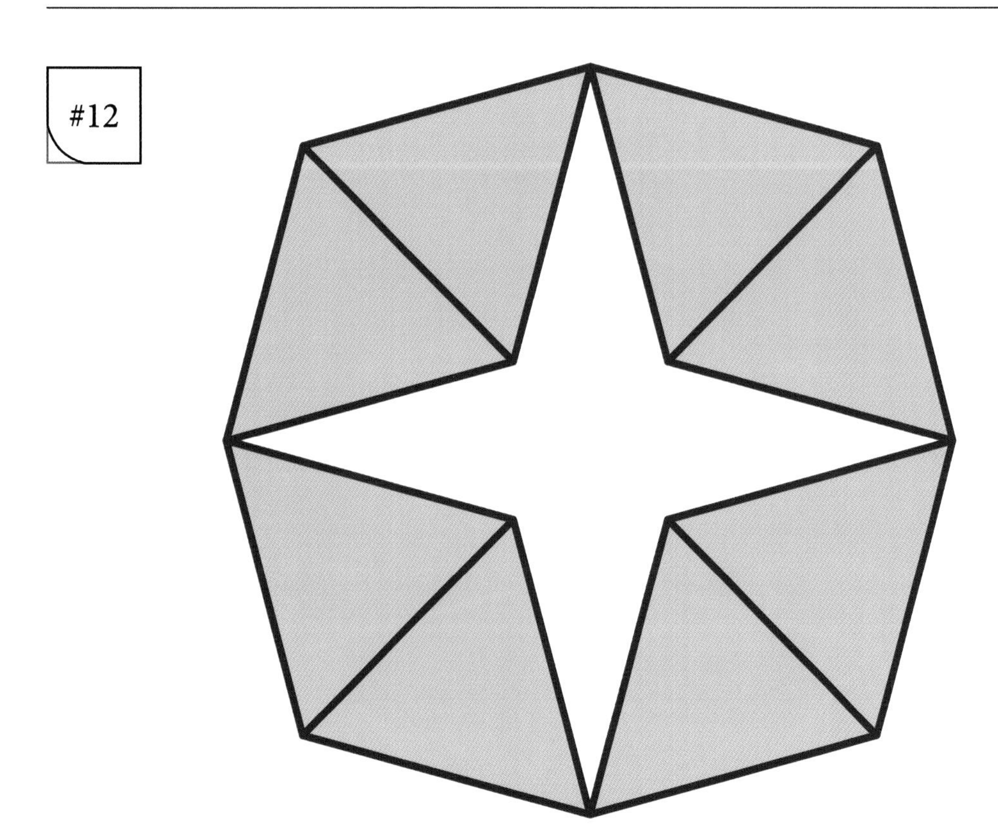

Eight congruent triangles are arranged around a square to form this octagon with 4 mirror lines.

What fraction of the octagon is shaded?

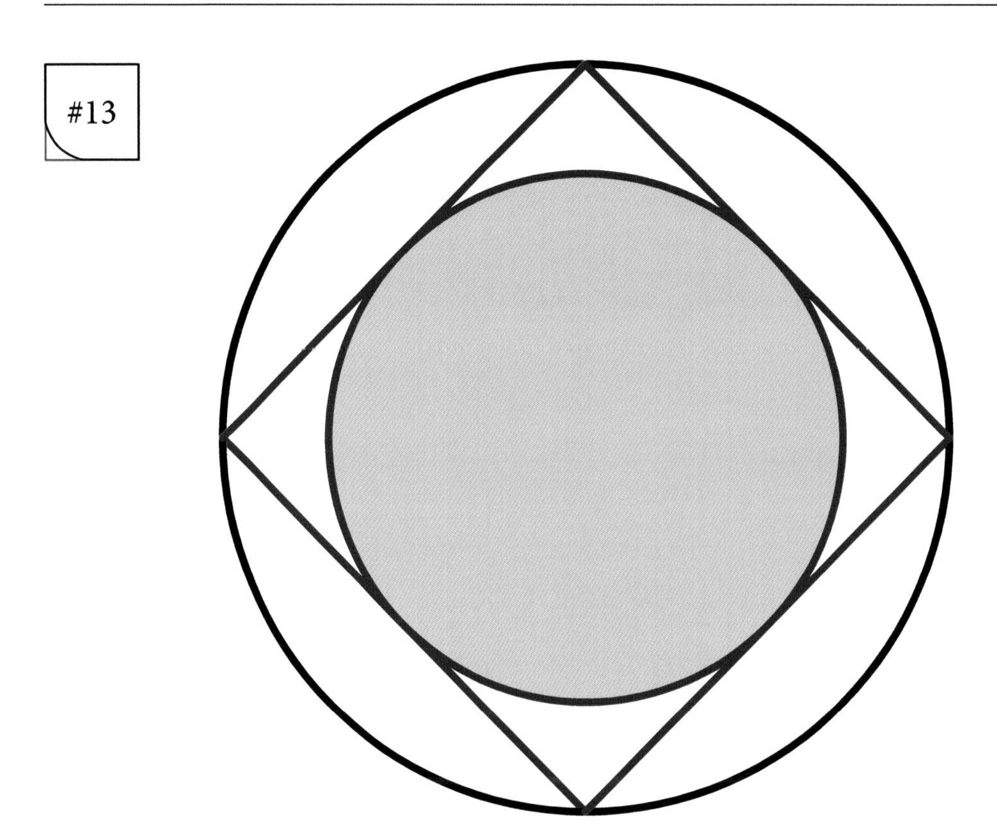

A square is shown with its inscribed and circumscribed circle.

What fraction of the larger circle is shaded?

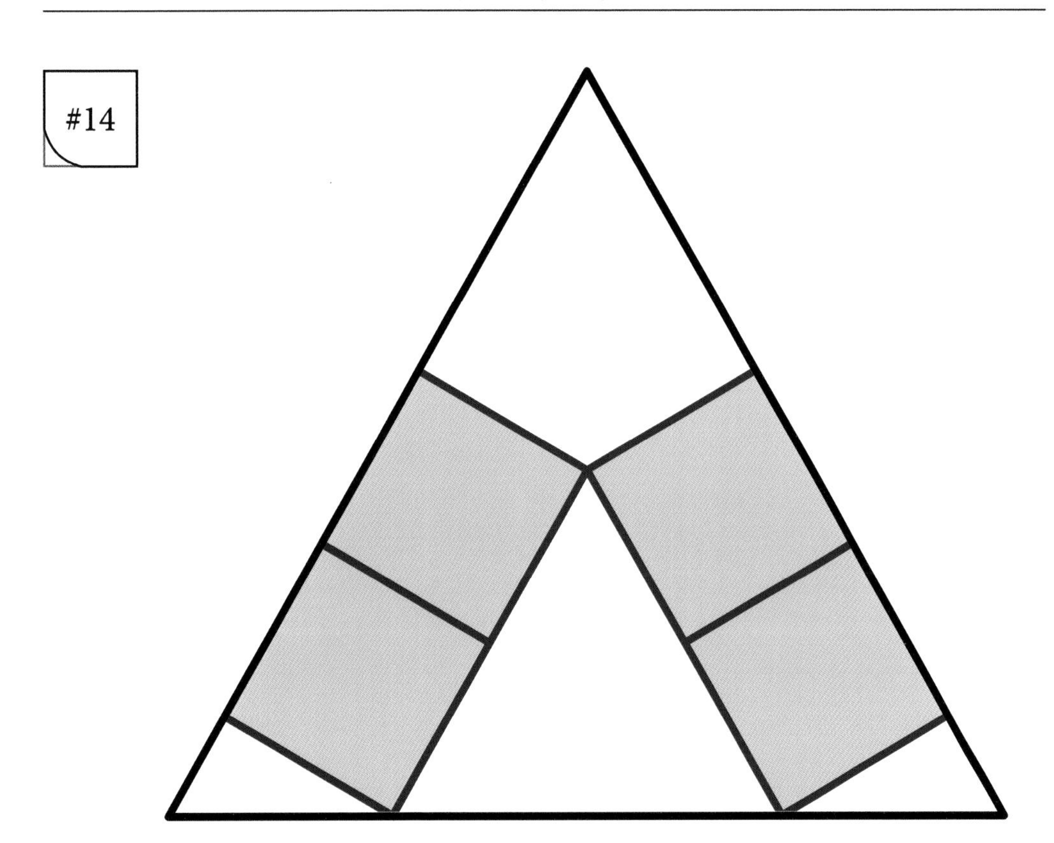

Four congruent squares are placed on the sides of an equilateral triangle so that the top two share a common vertex and the two bottom ones have a vertex on the base of the triangle.

What fraction of the equilateral triangle is shaded?

Solutions and Answers

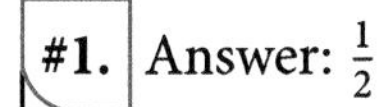

#1. Answer: $\frac{1}{2}$

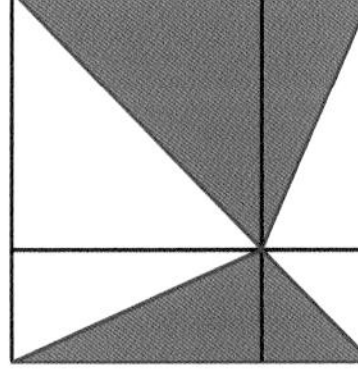

By segmenting the square as shown, you can see that half of each segment is shaded.

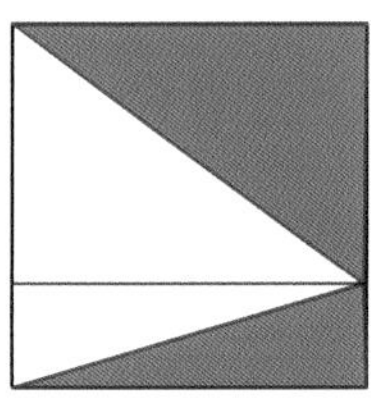

Using a parallel line to the square base, which passes through the point, we can shear the triangles along the line without changing their area (their base and height remain the same).

#2. Answer: $\frac{1}{3}$

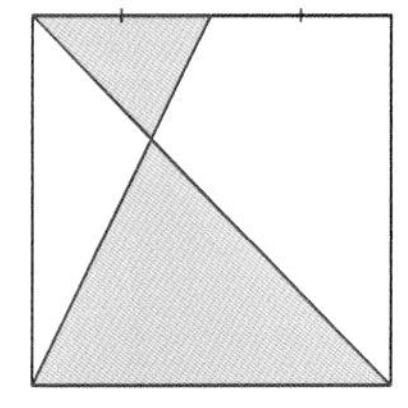

The two triangles highlighted are similar with a base ratio of $1:2$. Therefore their heights have the same ratio, meaning the height of the original shaded triangle is $\frac{2}{3}$ of the square side. Hence the answer must be $\frac{1}{2} \times \frac{2}{3} = \frac{1}{3}$.

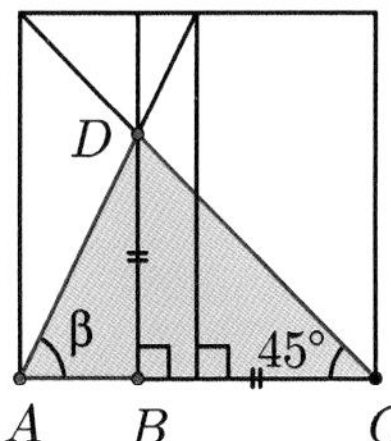

Assume the square has side 2. From the diagram, $DB = BC$, the area $\mathscr{A} = (BC + AB)(\frac{BC}{2})$. Using the vertical line from the midpoint of the bottom of the square, $\tan(\beta) = \frac{BC}{AB} = \frac{2}{1}$ therefore $2AB = BC$. Substitute this into our area formula: $\mathscr{A} = (2AB + AB) \times \frac{2AB}{2} = 3AB^2$. Hence the fraction shaded is $\frac{3AB^2}{(3AB)^2} = \frac{1}{3}$.

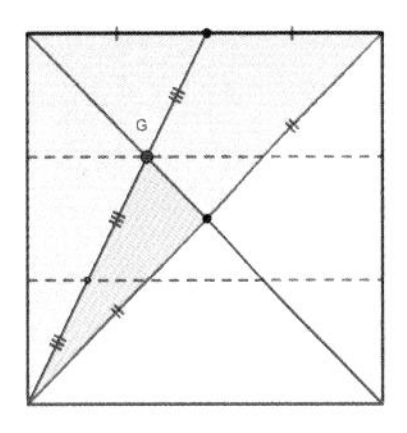

Point G is the centroid of the grey triangle which shows via Thales's theorem (aka the intercept theorem) that the altitude of our shaded triangle is two thirds of the square's side, so its area is a third of the square's area.

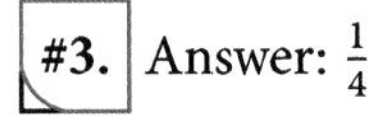

#3. Answer: $\frac{1}{4}$

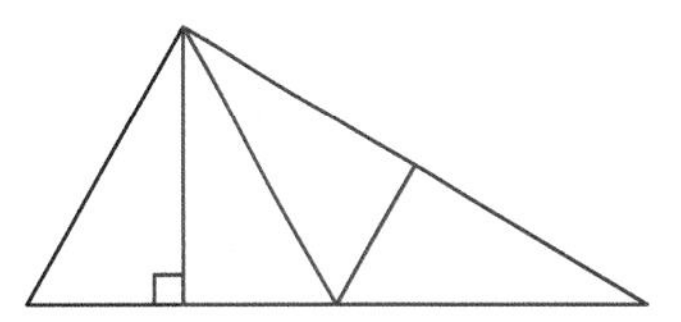

You can construct exactly four congruent triangles within the figure as shown.

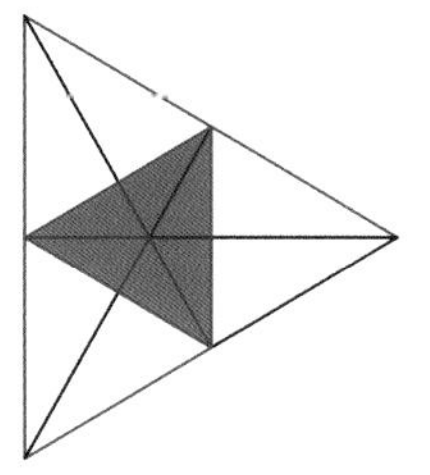

A series of six figures highlights the answer in the context of an equilateral triangle.

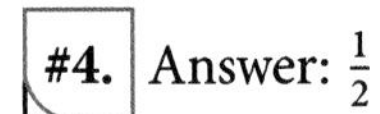

#4. Answer: $\frac{1}{2}$

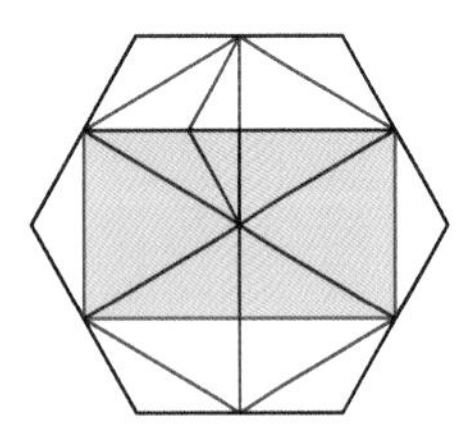

The additional lines demonstrate that the shaded region is equal to four equilateral triangles out of eight.

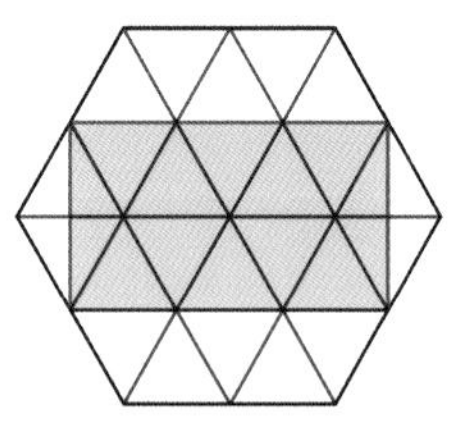

This alternative addition of lines shows that the shaded region is equal to 12 equilateral triangles out of 24.

#5. Answer: $\frac{1}{4}$

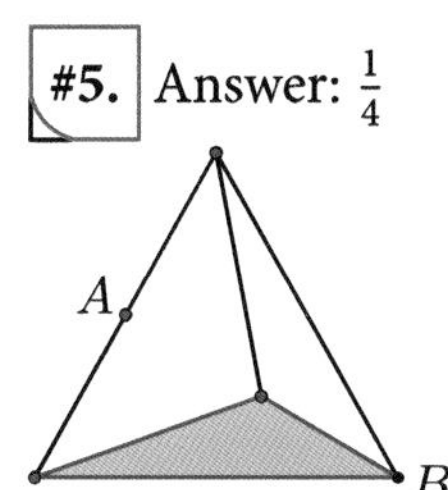

By adding in the segment shown, we can see that the largest white triangle is equal to half of the area of the equilateral triangle (it has half the height, and the same base). The remaining two triangles are congruent.

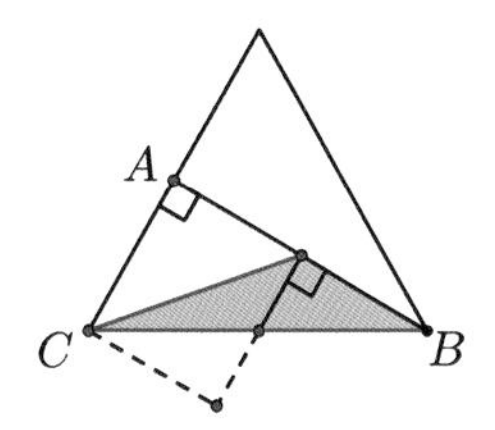

Drawing in a rectangle as shown, we can see that the shaded triangle takes up half of a half of it (it may be easier to visualise that the two white triangles in the rectangle take up three quarters of it).

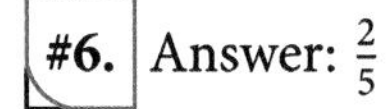

#6. Answer: $\frac{2}{5}$

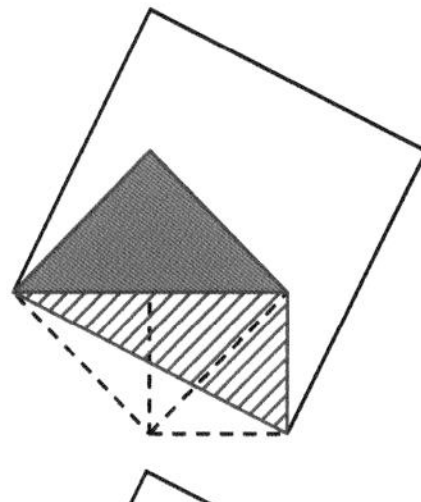

Assume the areas of the smaller squares are 1 and (therefore) 2, then the larger square would have area 5 using Pythagoras. The shaded area must be half of 2 (half the medium square) added to a 2 by 1 right triangle, which as a fraction is $\frac{2}{5}$.

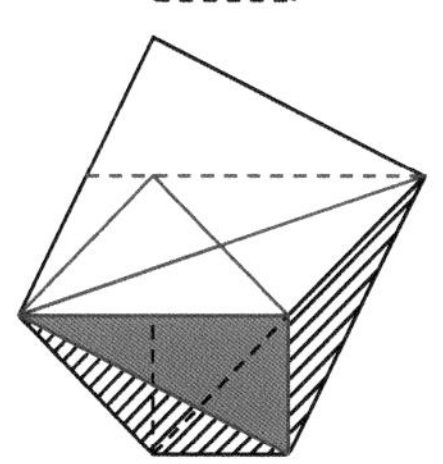

Shear the upper part of the original shade to show the total area is half the big square minus one of the two indicated hatched triangles – which have the same base/height as the blue shaded triangle. Assume measurements as with solution 1 to derive the answer.

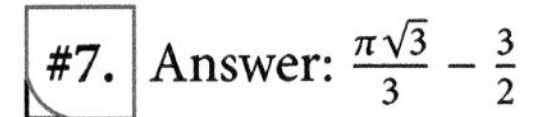

#7. Answer: $\frac{\pi\sqrt{3}}{3} - \frac{3}{2}$

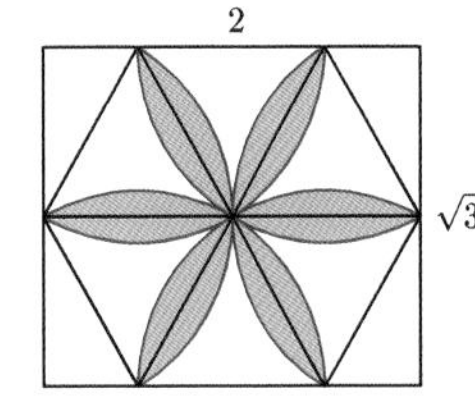

With the additional lines we can see that the rectangle has side lengths 2 and $\sqrt{3}$. We also have 12 unit circle segments with total area $12\left(\frac{\pi}{6} - \frac{\sqrt{3}}{4}\right) = 2\pi - 3\sqrt{3}$. Hence our shaded fraction is $\frac{2\pi-3\sqrt{3}}{2\sqrt{3}} = \frac{\pi\sqrt{3}}{3} - \frac{3}{2}$.

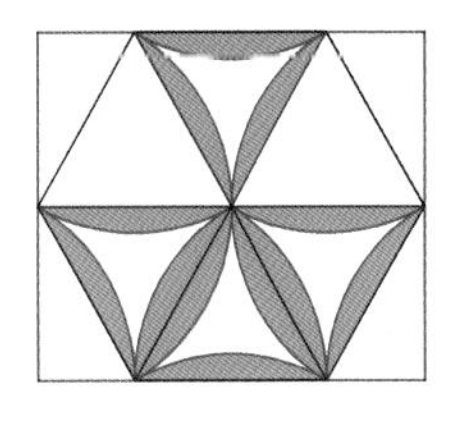

By rearranging the segments as shown, we can see that the total area is $2\sqrt{3}$ and the area not shaded is $\sqrt{3} + 4\left(\sqrt{3} - \frac{\pi}{2}\right)$. The white curvy triangle's area is explained with the second figure. So the numerator is $2\sqrt{3} - \sqrt{3} - 4\left(\sqrt{3} - \frac{\pi}{2}\right) = 2\pi - 3\sqrt{3}$ and the denominator is $2\sqrt{3}$. Simplifying is left to the reader.

#8. Answer: $\frac{1}{2}$

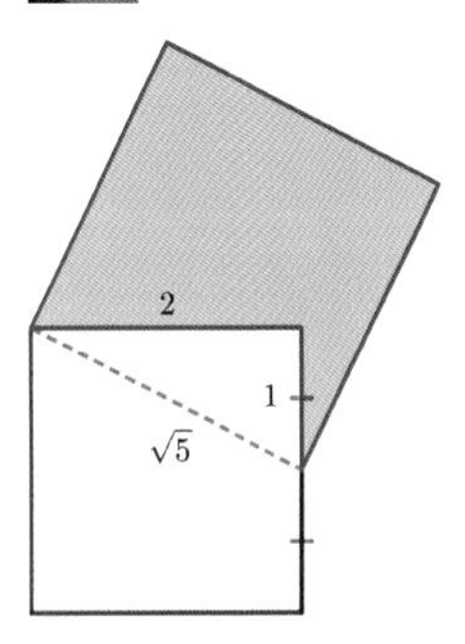

Assign a length of 2 (area 4) to the smaller square, then the larger square has area 5 using Pythagoras. Therefore the total area of the hexagon is $4 + (5 - 1) = 8$ and so the shaded fraction must be $\frac{4}{8}$.

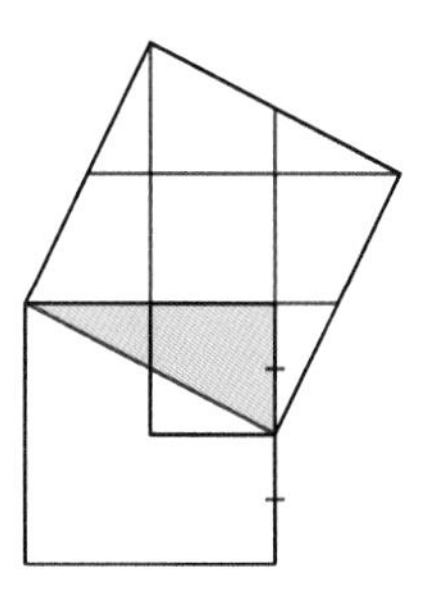

Drawing in the lines indicated, you can see that we can rearrange each small triangle in the largest square to reconstruct it into five squares in a cross shape. Hence the shaded triangle is a fifth of the large square, and equal to a quarter of the small square. Therefore four fifths of the largest square are shaded in the problem, which are equal in area to four quarters of the other square, so half is shaded.

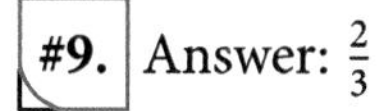

#9. Answer: $\frac{2}{3}$

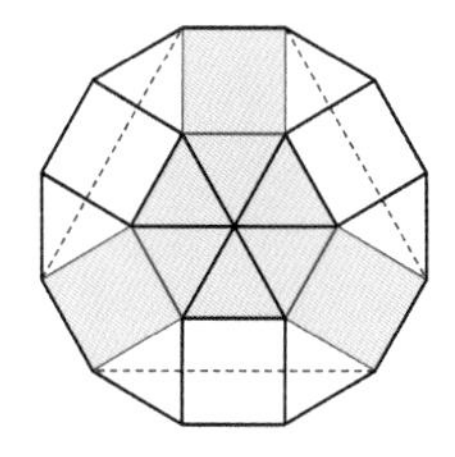

• By drawing three more squares, one sees this regular dodecagon. The shaded figure represents $6T + 3S$ (where T is the area of an equilateral triangle and S that of a square). To get the whole figure you must add $x = 3T + \frac{3}{2}S$ i.e. a half of the shaded area. $\frac{2x}{2x+x} = \frac{2}{3}$.

•• The white area in the dodecagon represents a half of the area D of the dodecagon, so is the shaded region. So the total shape has an area of $D - \frac{D}{4} = \frac{3D}{4}$. The sought fraction is $\frac{D/2}{3D/4} = \frac{1}{2} \times \frac{4}{3} = \frac{2}{3}$.

#10. Answer: $\frac{1}{3}$

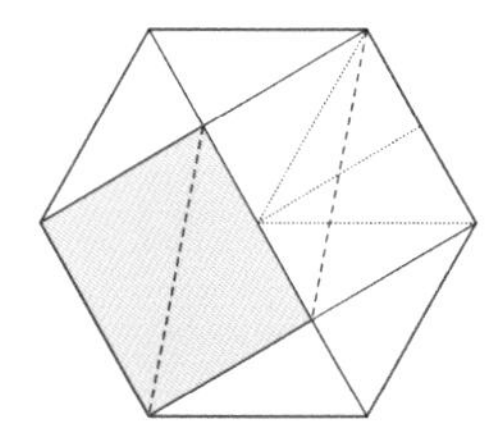

By shearing the parallelogram, or translating half of it, you see its area is that of this rectangle. Using the disection with halves of equilateral triangles, the shaded area appears to represent the area of two of the six equilateral triangles in the regular hexagon.

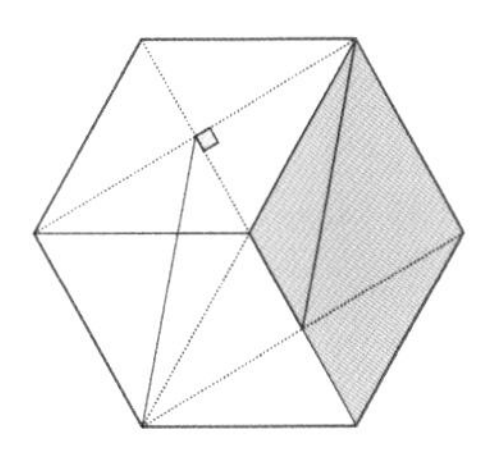

The parallelogram has a side measuring half a diagonal of the shaded rhombus, and a height equal to twice the other diagonal of the rhombus. You can also see that both figures have the same area using two shearings. Three of these rhombuses form the hexagon, which gives the answer of $\frac{1}{3}$.

#11. Answer: $\frac{7}{24}$

The only tricky part is the top left quarter of this square and its highlighted triangle.

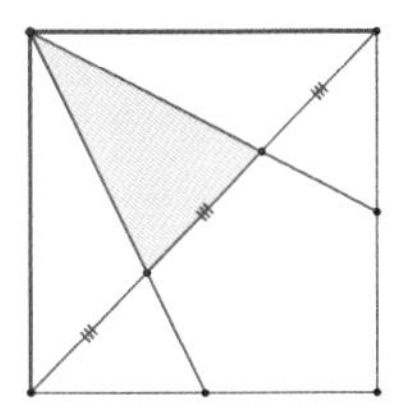

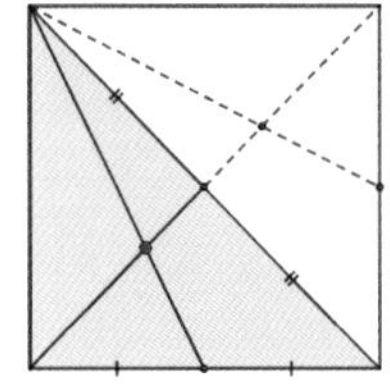

Consider the shaded right isosceles triangle, the big point is its centroid, so it cuts the median (half the diagonal of the square) in thirds. By symmetry we see that the construction produces a trisection of the diagonal of this square. The shaded triangle in the top left quarter of the figure represents $\frac{1}{4} \times \frac{1}{3} \times \frac{1}{2} = \frac{1}{24}$. The three remaining right triangles add up to a quarter of the square. $\frac{1}{24} + \frac{1}{4} = \frac{7}{24}$.

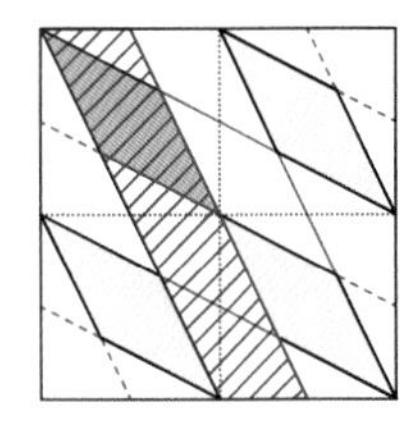

By reproducing the same pattern with the rhombus in each quarter of the square as shown on the figure, you'll see that its area is a third of the area of the hatched parallelogram, so it is $\frac{1}{3}\left(\frac{1}{4} \times 1\right) = \frac{1}{12}$. Divide it by two to find that the area of the top most triangle is $\frac{1}{24}$. The final calculation is then identical.

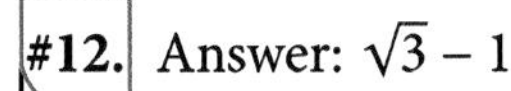

#12. Answer: $\sqrt{3} - 1$

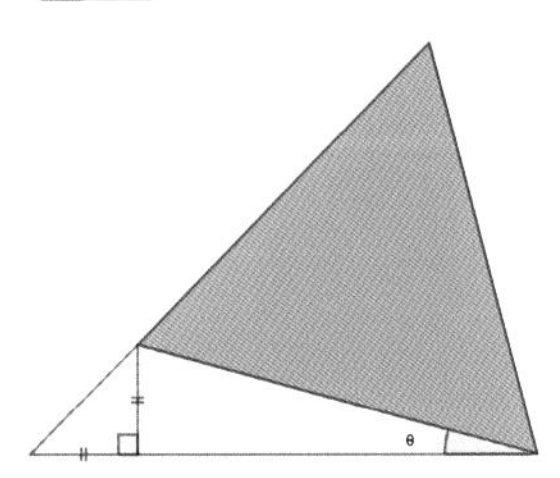

Is is sufficient to solve the problem in this eighth of the figure. The marked angle θ measures 15° (It is complementary to 180 – 45 – 60 = 75). If we choose the side of the equilateral triangle as unit, the white triangle has a height of $\sin(\theta)$ so its area is $\frac{1}{2}\sin(\theta) \times (\cos(\theta) + \sin(\theta))$. Using $\sin(\theta)\cos(\theta) = \frac{1}{2}\sin(2\theta)$ and $\sin^2(\theta) = \frac{1}{2}(1 - \cos(2\theta))$ we get:

$$\frac{1}{4}(\sin(2\theta) + 1 - \cos(2\theta)) = \frac{1}{4}\left(\frac{1}{2} + 1 - \frac{\sqrt{3}}{2}\right) = \frac{3 - \sqrt{3}}{8}$$

The area of the equilateral triangle being $\frac{\sqrt{3}}{4}$, the total area is $\frac{3+\sqrt{3}}{8}$ and the shaded region represents $\frac{2\sqrt{3}}{3+\sqrt{3}} = \sqrt{3} - 1$.

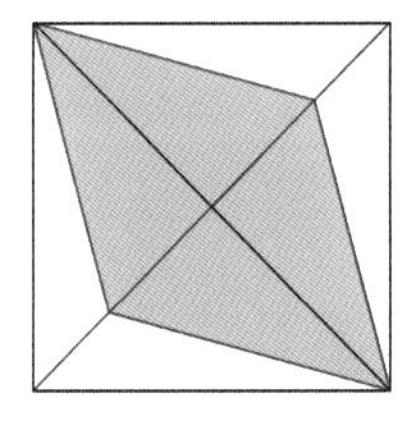

Using a quarter of the figure and completing the square, the equilateral triangle appears to have a height equal to half the diagonal of this square. For a unit side length of the equilateral triangle the square has a diagonal of $\sqrt{3}$ so its side is $\frac{\sqrt{3}}{\sqrt{2}} = \frac{\sqrt{6}}{2}$. The white area is therefore $\frac{6}{4} - \frac{\sqrt{3}}{2} = \frac{3-\sqrt{3}}{2}$ and the fraction sought is $\frac{\sqrt{3}}{2} \div \left(\frac{\sqrt{3}}{2} + \frac{3-\sqrt{3}}{4}\right) = \sqrt{3} - 1$.

#13. Answer: $\frac{1}{2}$

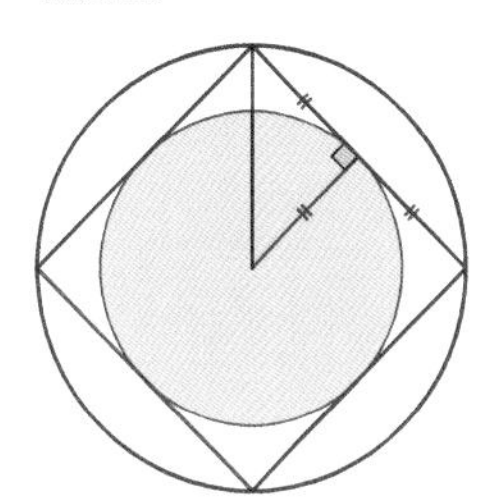

A radius R of the big circle, half a side and a radius r of the small circle form a right isosceles triangle. Therefore $R = r\sqrt{2}$ so $r^2 = \frac{1}{2}R^2$ and the shaded area represents half of the big circle's area.

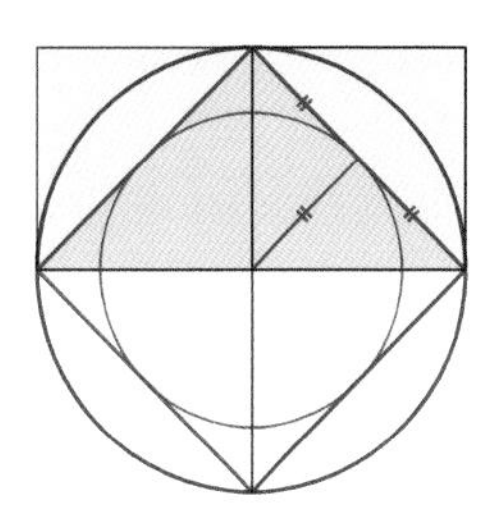

The area of the square can be calculated in two ways: Using the small radius r it is $(2r)^2 = 4r^2$. Using the big radius R which is half a diagonal you see it is $2 \times R^2$. So $2R^2 = 4r^2$ and again $r^2 = \frac{1}{2}R^2$.

#14. Answer: $28\sqrt{3} - 48$

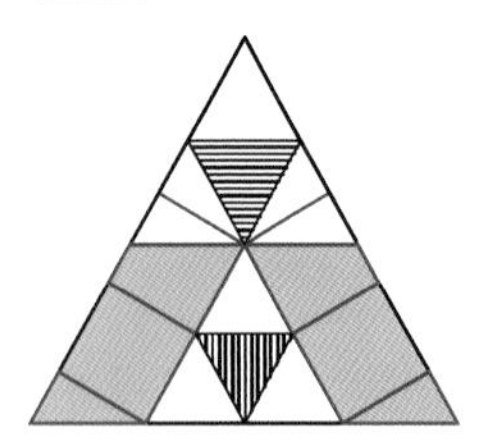

This dissection shows that the total area can be dissected into four unit squares, four equilateral triangles with unit side length and four with side length $\frac{2}{\sqrt{3}}$. An equilateral triangle's area with sides of a is $\frac{\sqrt{3}}{4}a^2$ so the sought fraction is the reciprocal of $1 + \frac{\sqrt{3}}{4}\left(1 + \frac{4}{3}\right) = 1 + \frac{7\sqrt{3}}{12}$ which is $28\sqrt{3} - 48 \approx 0.497$.

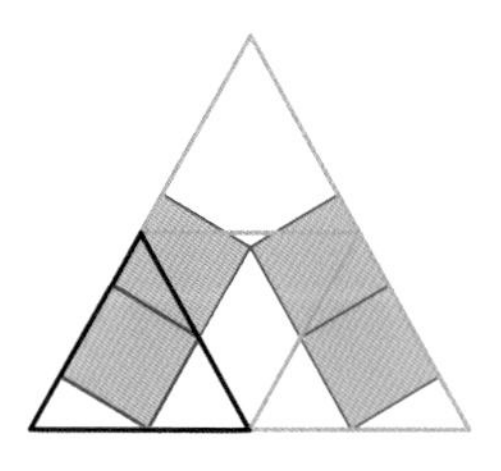

Joining the midpoints of the equilateral triangle forms four congruent equilateral triangles so the fraction sought is identical to the one we would have for one square in its circumscribing equilateral highlighted at the bottom left. Taking a unit square, since $\cos(30) = \sqrt{3}/2$, the side of the equilateral triangle is $1 + \frac{2}{\sqrt{3}}$.

The fraction sought is the reciprocal of $\frac{\sqrt{3}}{4}\left(1 + \frac{2}{\sqrt{3}}\right)^2 = 1 + \frac{7\sqrt{3}}{12}$.

What’s the Angle?

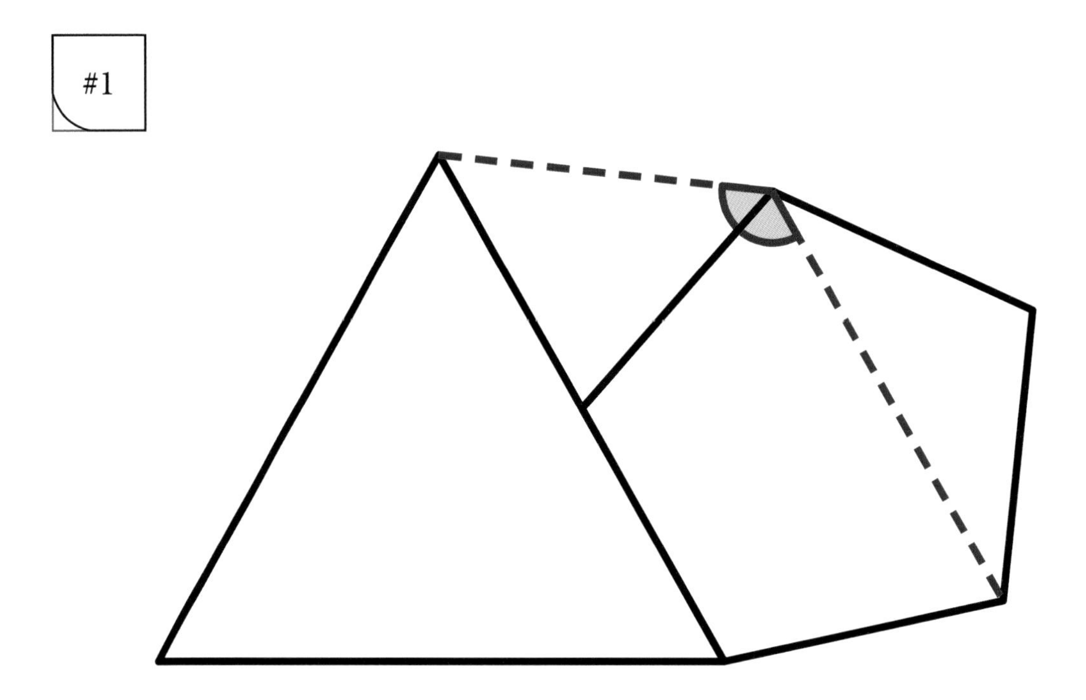

A regular pentagon is constructed from the midpoint of the side of an equilateral triangle as shown.

Find the missing angle.

#2

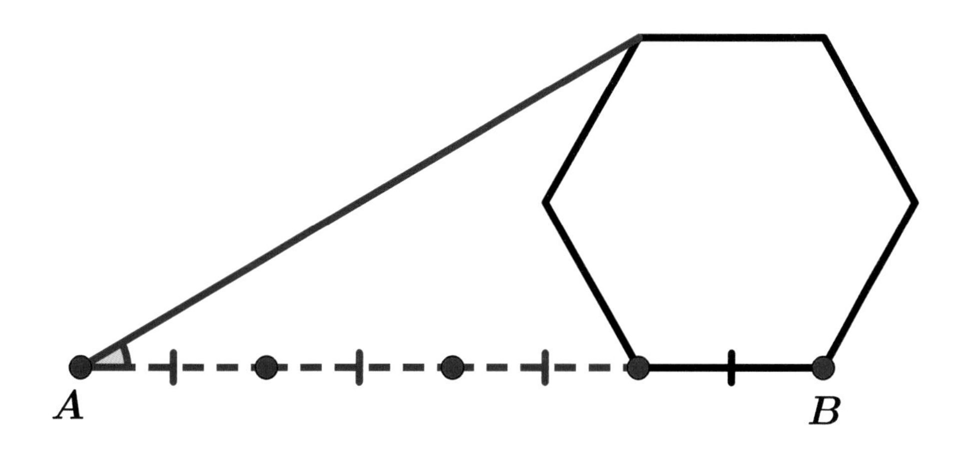

AB is a straight line leading to the construction of a regular hexagon.

Find the missing angle.

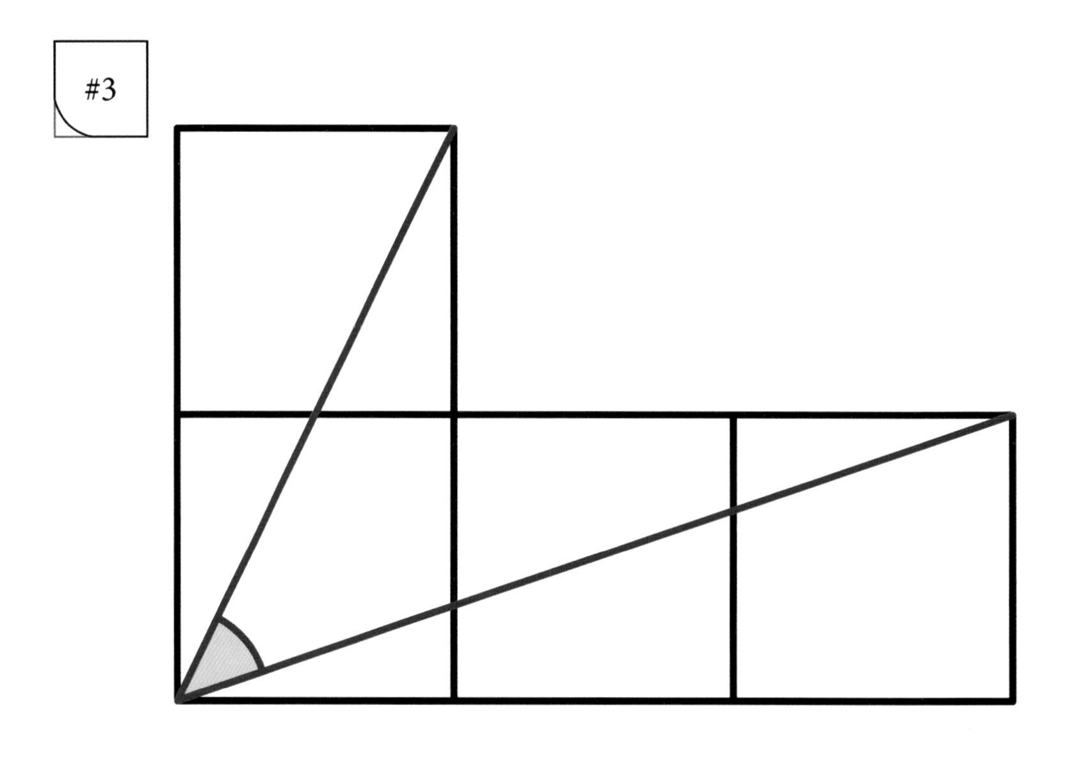

Four congruent squares are constructed as a tetromino as shown.

Find the missing angle.

#4

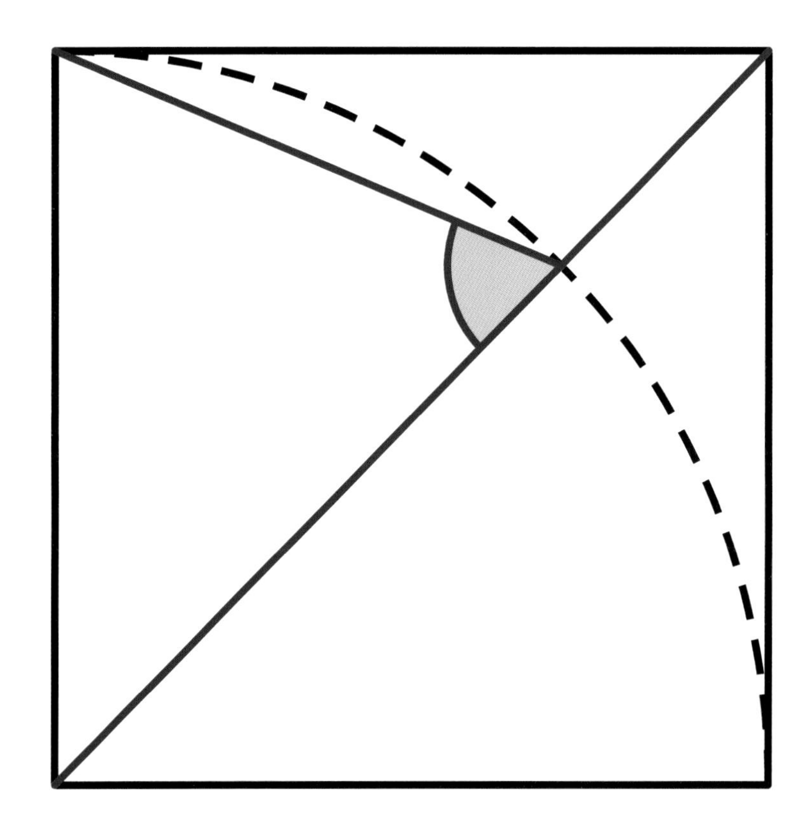

A square and a circle sector are constructed as shown.

Find the missing angle.

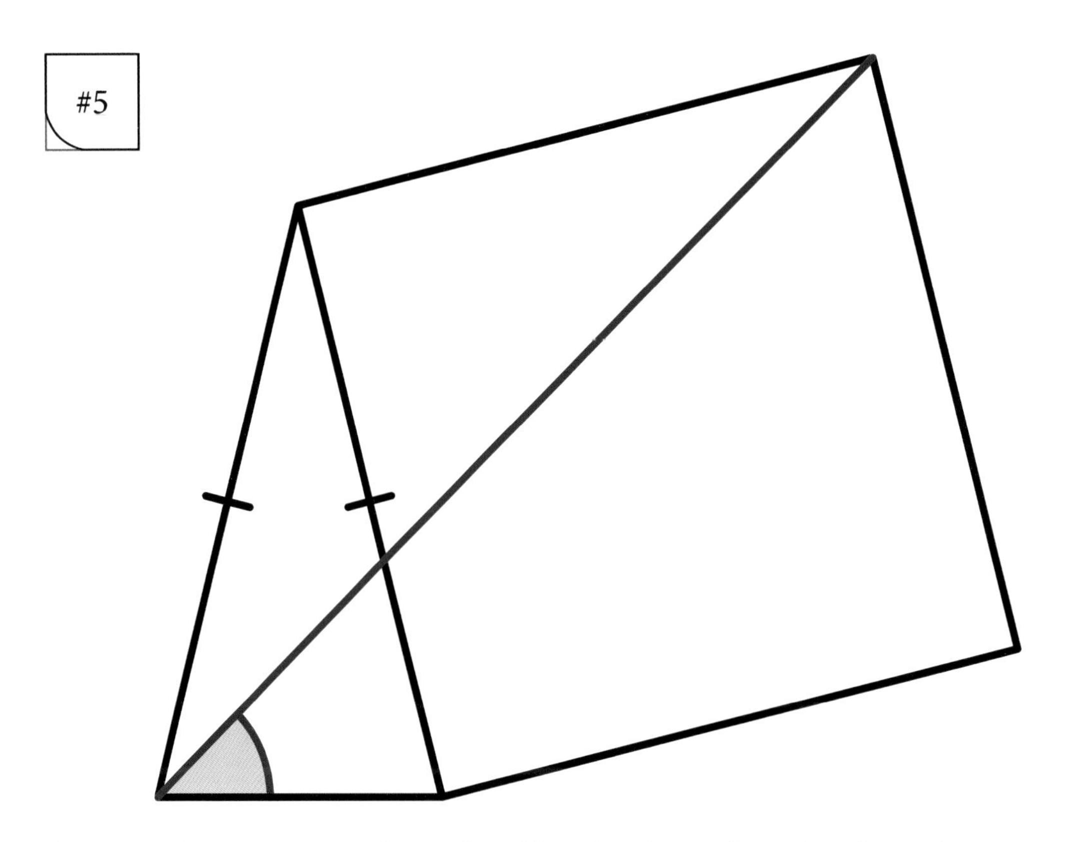

A square is constructed from the side of an isosceles triangle as shown.

Find the missing angle.

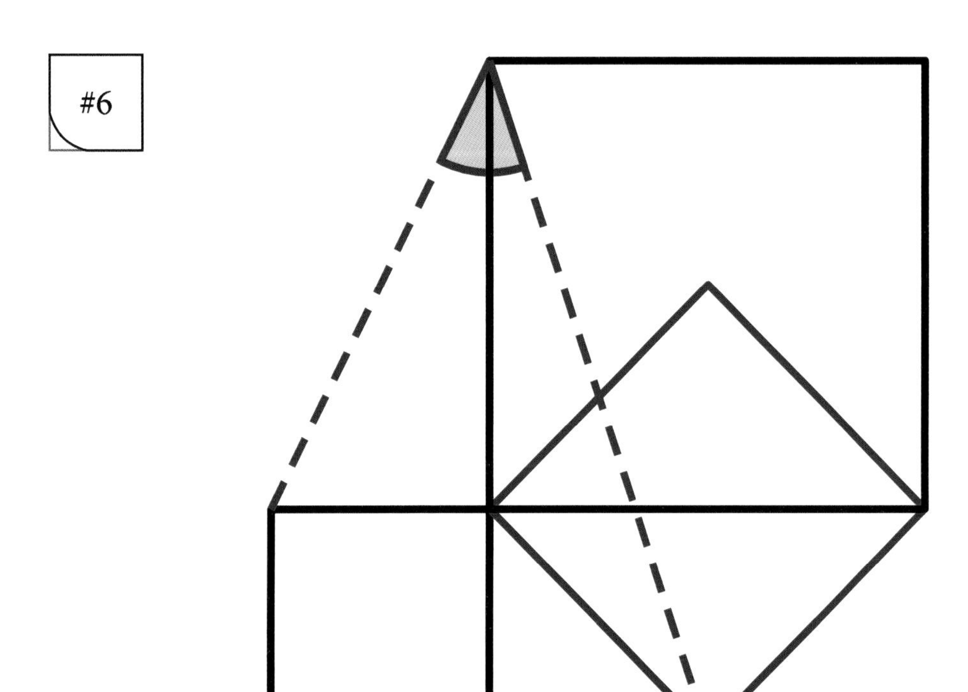

Two black squares are constructed by drawing six straight lines, a third square is constructed such that *A*, *B*, *C* are collinear.

Find the missing angle.

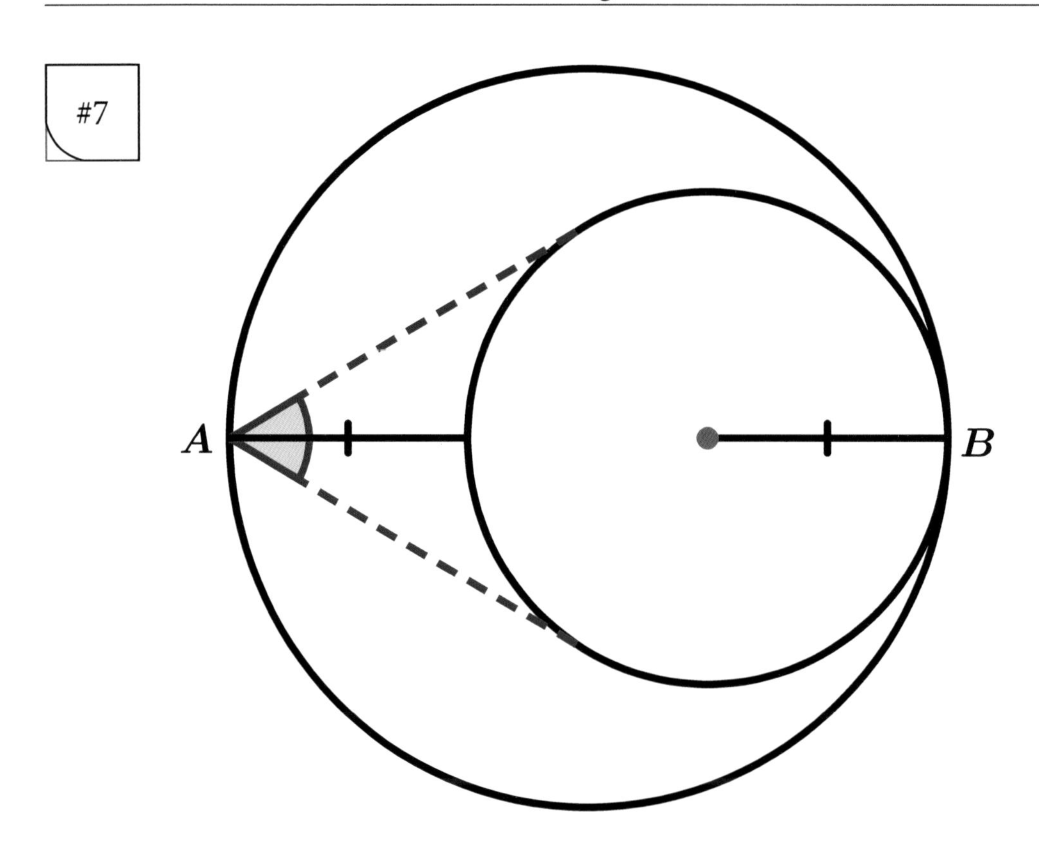

Two cotangent circles are constructed such that *AB* is a straight line as shown. The dashed lines are tangent to the smaller circle.

Find the missing angle.

#8

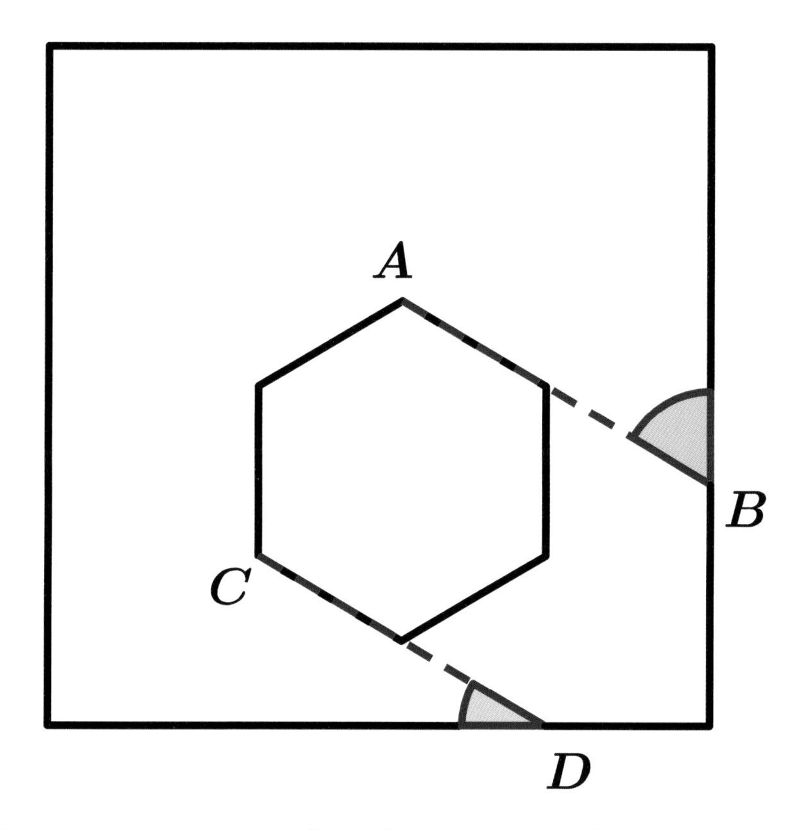

A regular hexagon is constructed inside a square as shown.

Find the sum of the two highlighted angles. Assume *AB* and *CD* are straight lines.

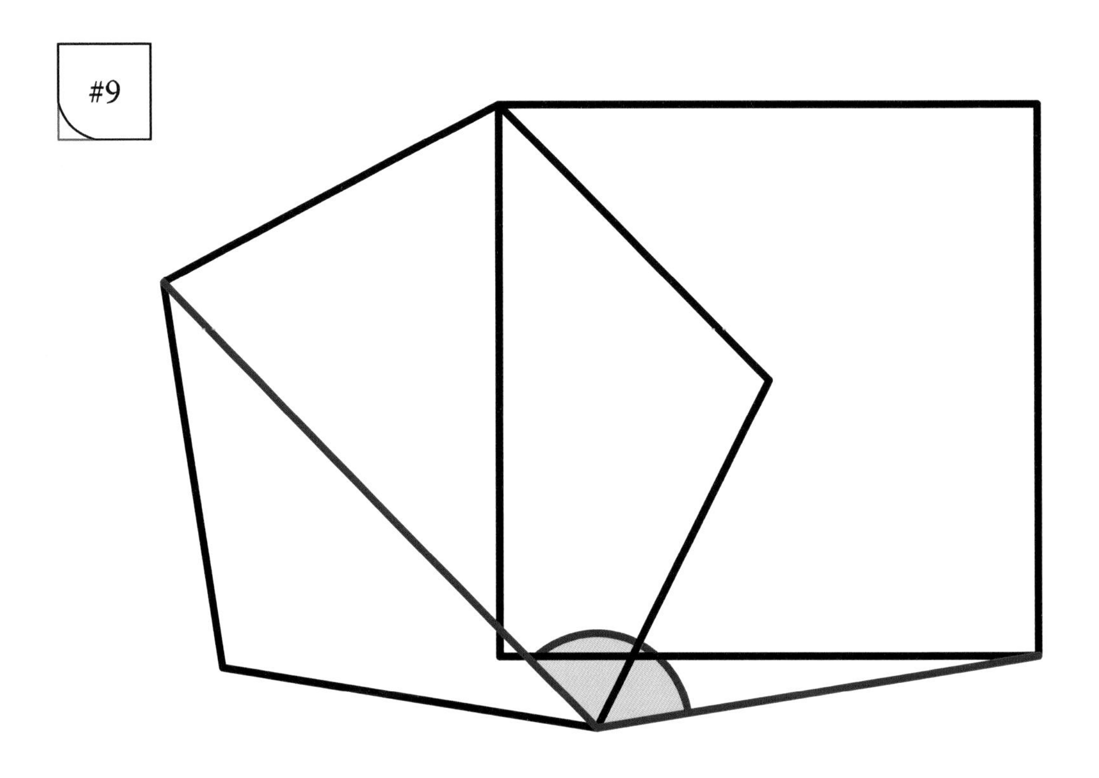

A regular pentagon is constructed from the centre point and a vertex of a square as shown.

Find the missing angle.

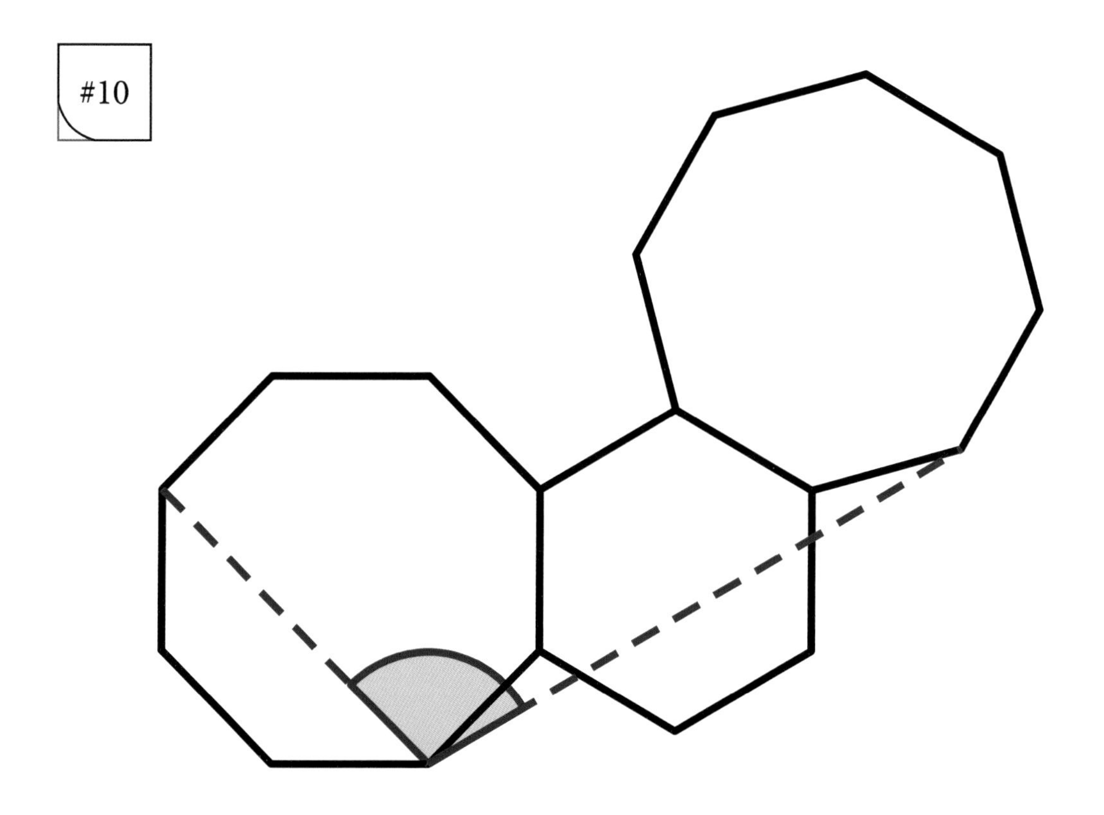

Two regular octagons are constructed on the sides of a regular hexagon as shown.

Find the missing angle.

Solutions and Answers

#1. Answer: 126°

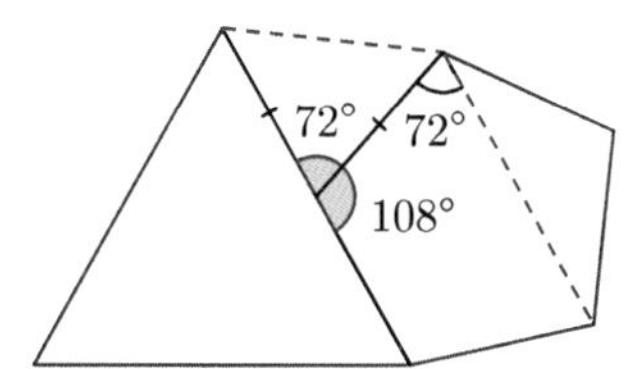

Using the interior angle of the pentagon, we can deduce the angle between the pentagon and triangle is 72°. Therefore each of the other two angles in the isosceles triangle made by the gap between the two shapes are 54°. 54°added to the 72° inside the pentagon gives 126°.

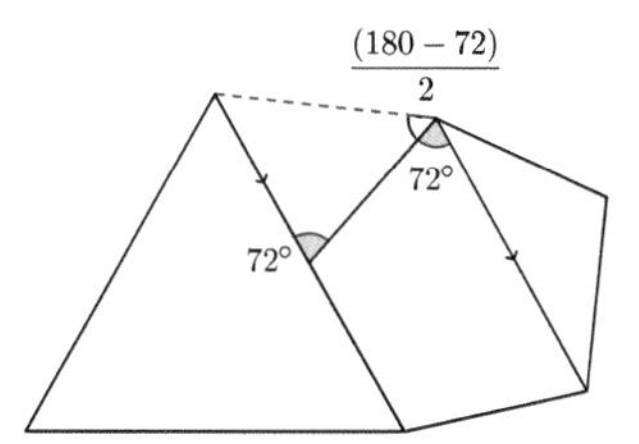

By identifying the parallel lines shown, you can deduce that both indicated angles must be 72° hence the missing part of the required angle is $\frac{180-72}{2}$.

#2. Answer: 30°

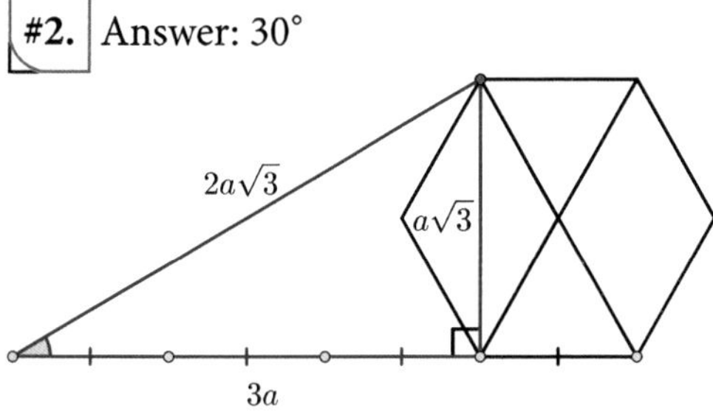

Let the right angled triangle shown have base $3a$, and therefore height $a\sqrt{3}$ (twice the height of the equilateral triangle). By Pythagoras, the hypotenuse would therefore be $2a\sqrt{3}$. This gives side ratios of $1 : 2 : \sqrt{3}$ which is the side ratio of a 30°, 60°, 90° triangle.

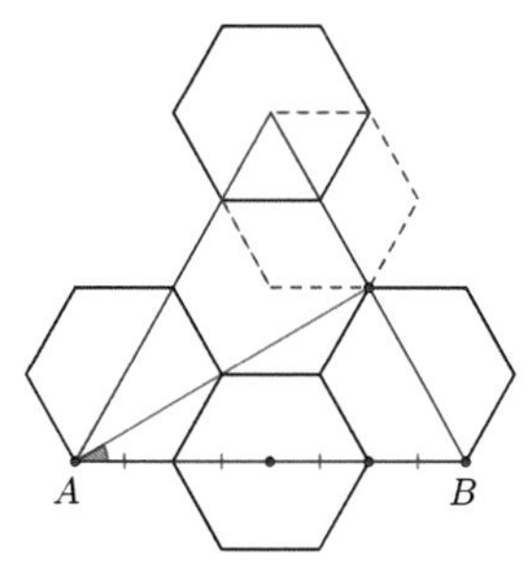

By tessellating hexagons, you can visualise a large equilateral triangle and see that the angle required is from a vertex to an opposite midpoint.

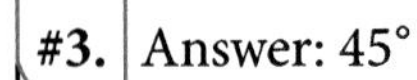

#3. Answer: 45°

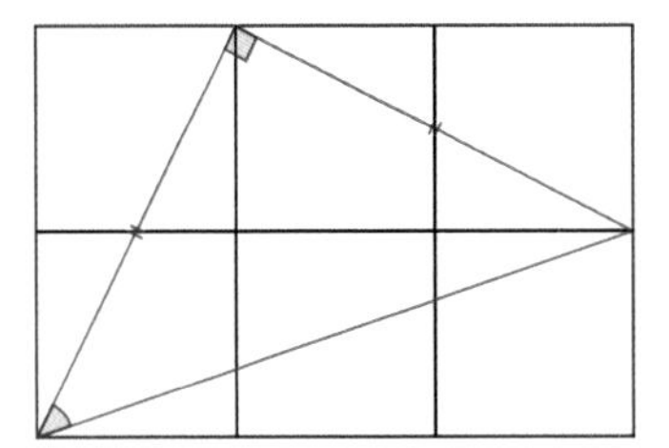

Proof without words.

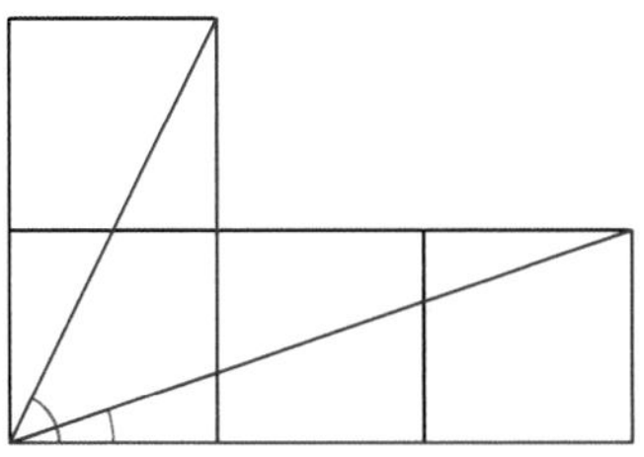

The angle is $\alpha = \tan^{-1}(2) - \tan^{-1}(\frac{1}{3})$. The formula $\tan(x - y) = \frac{\tan(x) - \tan(y)}{1 + \tan(x)\tan(y)}$ gives:

$$\tan(\alpha) = \frac{2 - \frac{1}{3}}{1 + 2 \times \frac{1}{3}} = \frac{6 - 1}{3 + 2} = 1$$

$\tan(\alpha) = 1$ so $\alpha = 45°$.

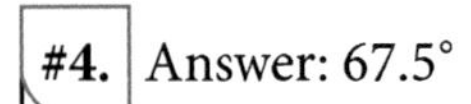

#4. Answer: 67.5°

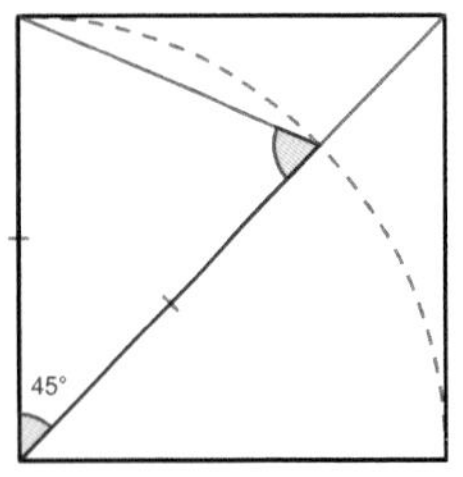

$$\frac{180 - 45}{2} = 67.5$$

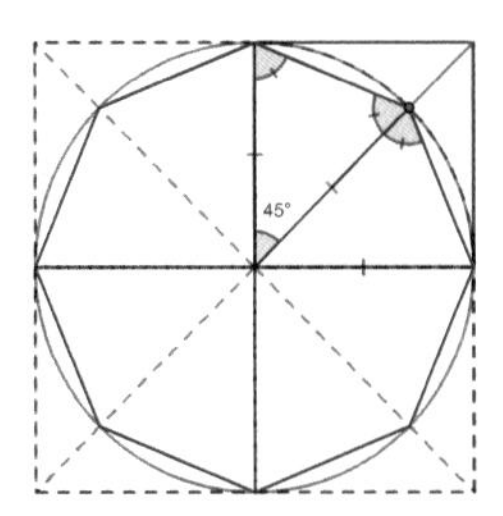

By tessellating four copies of the shape, we can see the formation of a regular octagon.

#5. Answer: 45°

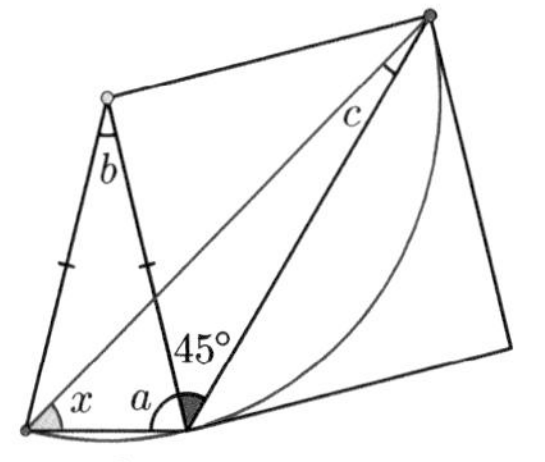

$c = \frac{b}{2}$, $x = 180 - (a + 45 + c) =$
$180 - (a + \frac{b}{2}) - 45 = 90 - 45 = 45$

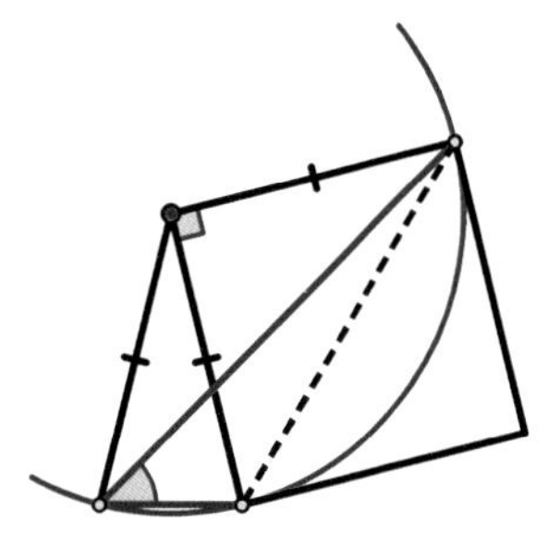

Consider the dashed chord of the circle, the angle sought is half the central right angle.

#6. Answer: 45°

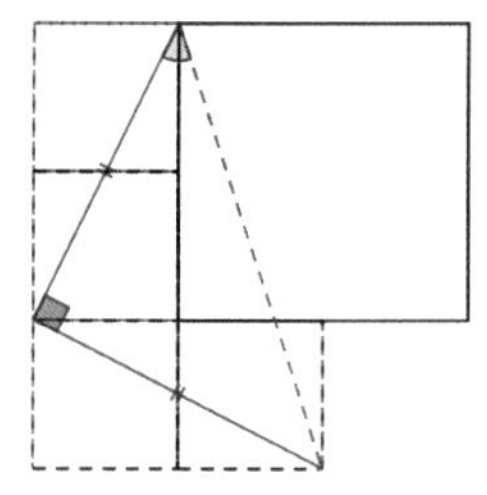

This is the same situation as snack #3

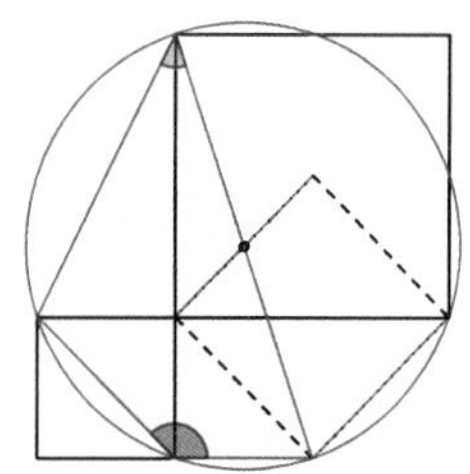

Construct a cyclic quadrilateral as shown. Opposite angles sum to 180° and the angle at the bottom must be 135°.

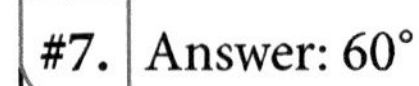

#7. Answer: 60°

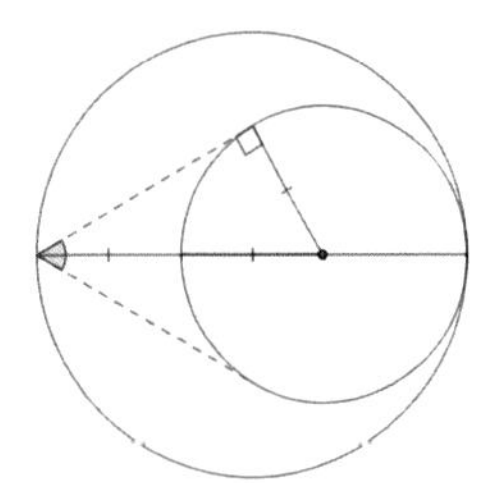

The right angled triangle indicated has a side ratio of $1 : 2 : \sqrt{3}$ like in snack #2. Hence half of the required angle is 30°.

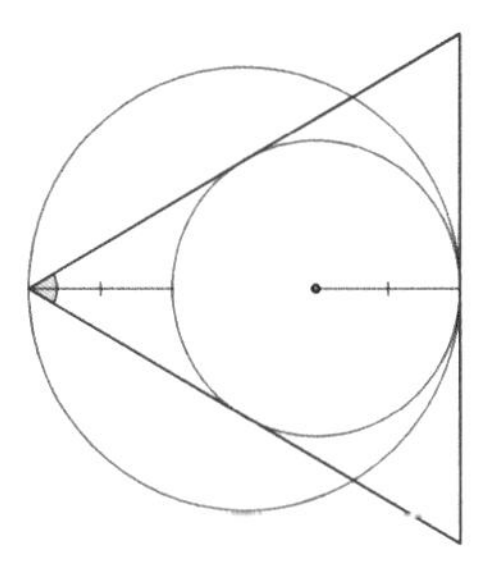

The small circle is the incircle of the blue triangle, and its centre is at two thirds of a median, so it is also the centroid. Hence, the blue triangle is equilateral.

#8. Answer: 90°

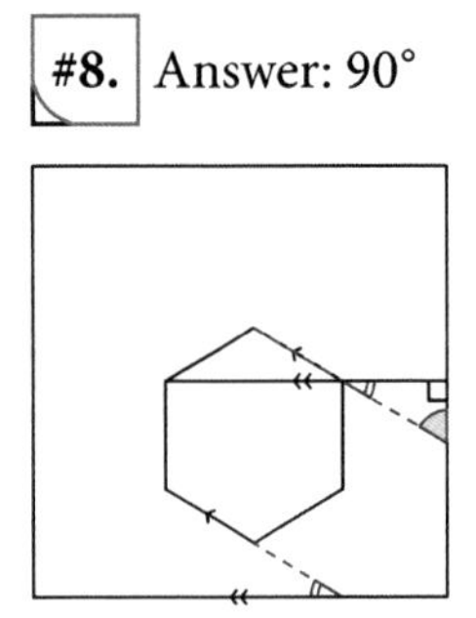

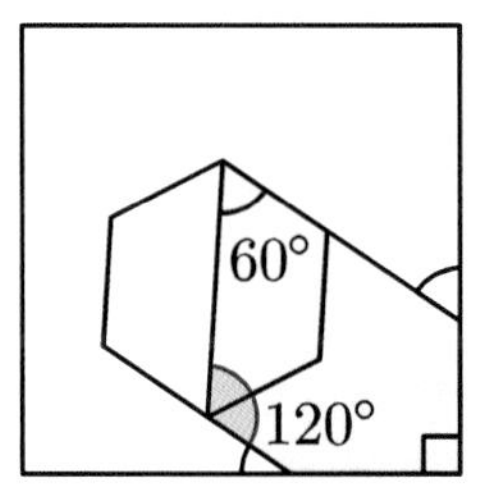

Two proofs without words.

#9. Answer: 126°

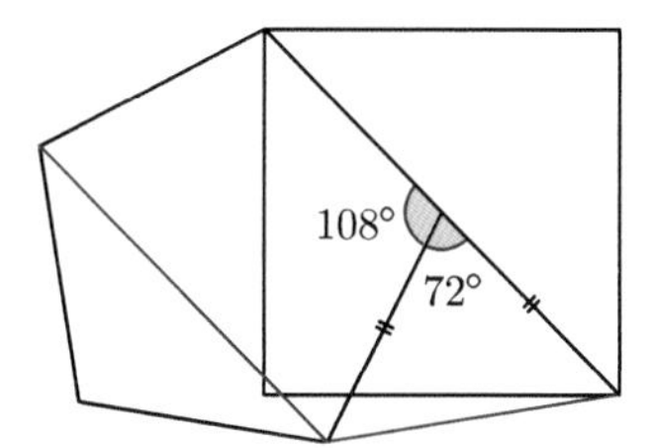

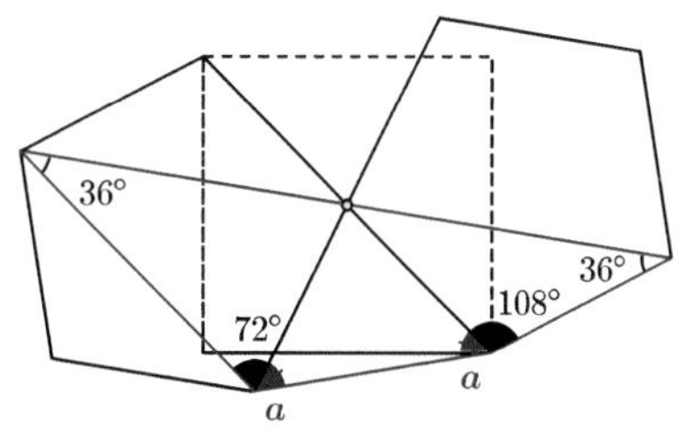

An isosceles triangle can be derived from the fact that all sides of the pentagon are equal in length, and that one side is equal to half the diagonal of the square. Using the internal angle of a pentagon, the internal angles of the isosceles triangle can be derived and then the solution itself.

By constructing a congruent pentagon on the other side of the square, an irregular quadrilateral can be seen. All angles can be derived from the internal angles of regular pentagons, except for the two congruent angles a indicated in the diagram.
As angles in a quadrilateral sum to 360°:
$a = \frac{360-(36\times2+72+108)}{2} = 54.$

#10. Answer: 105°

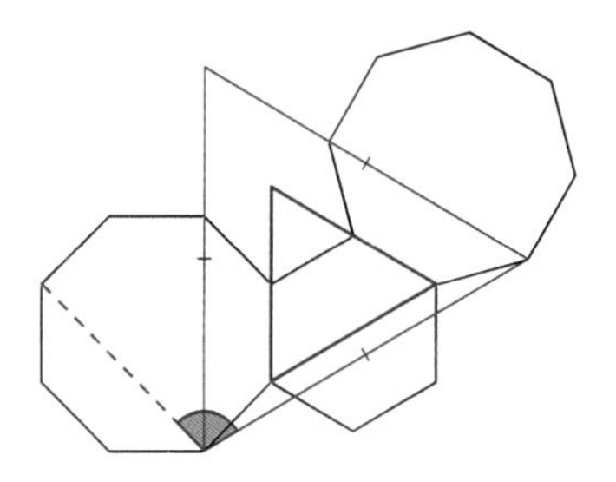

The smaller equilateral triangle can be constructed by extending the lines from the hexagon. The larger similar triangle can be constructed using the fact that joining vertices creates parallel lines to the smaller triangle. This allows us to derive the size of the part of the missing angle that falls outside of the shape (60°). The rest of the angle is two sixths of the internal angle: $\frac{2}{6} \times 135 = 45$. You can also see it with the central angle (90°) of the octagon.

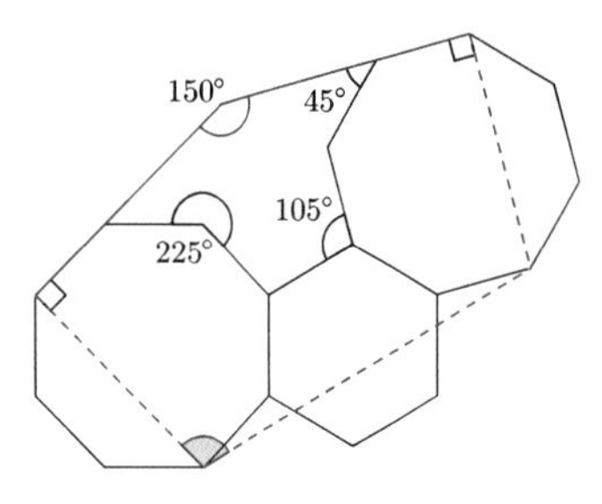

Joining vertices of the two octagons as shown, you can deduce the angle of 150° using the irregular heptagon formed outside of the three shapes, and the external angles of the original shapes. The large irregular pentagon formed now has only two congruent missing angles, one of which is the angle required to solve the problem (the other two are right angles). As the sum of internal angles of a pentagon is 540°, the rest is left to the reader.

Sangaku

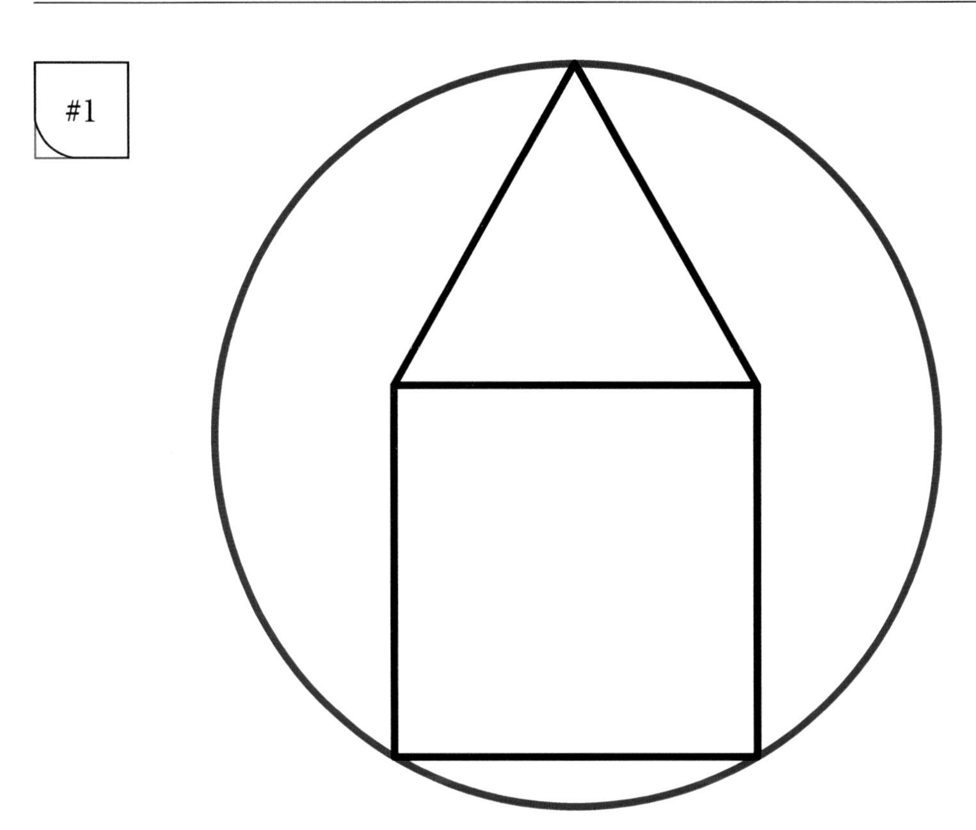

A unit square and an equilateral triangle are constructed within a circle as shown.

What is the radius of the circle?

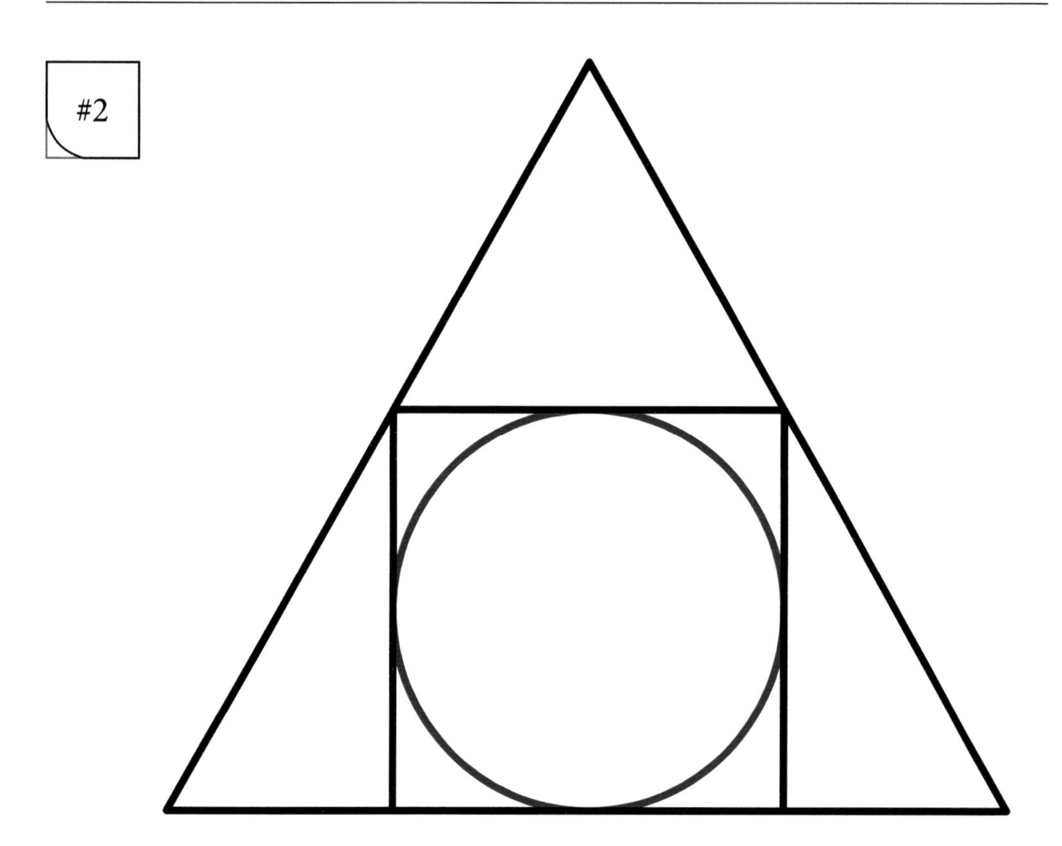

A unit circle is inscribed in a square, which in turn is used to construct an equilateral triangle as shown.

What is the area of the triangle?

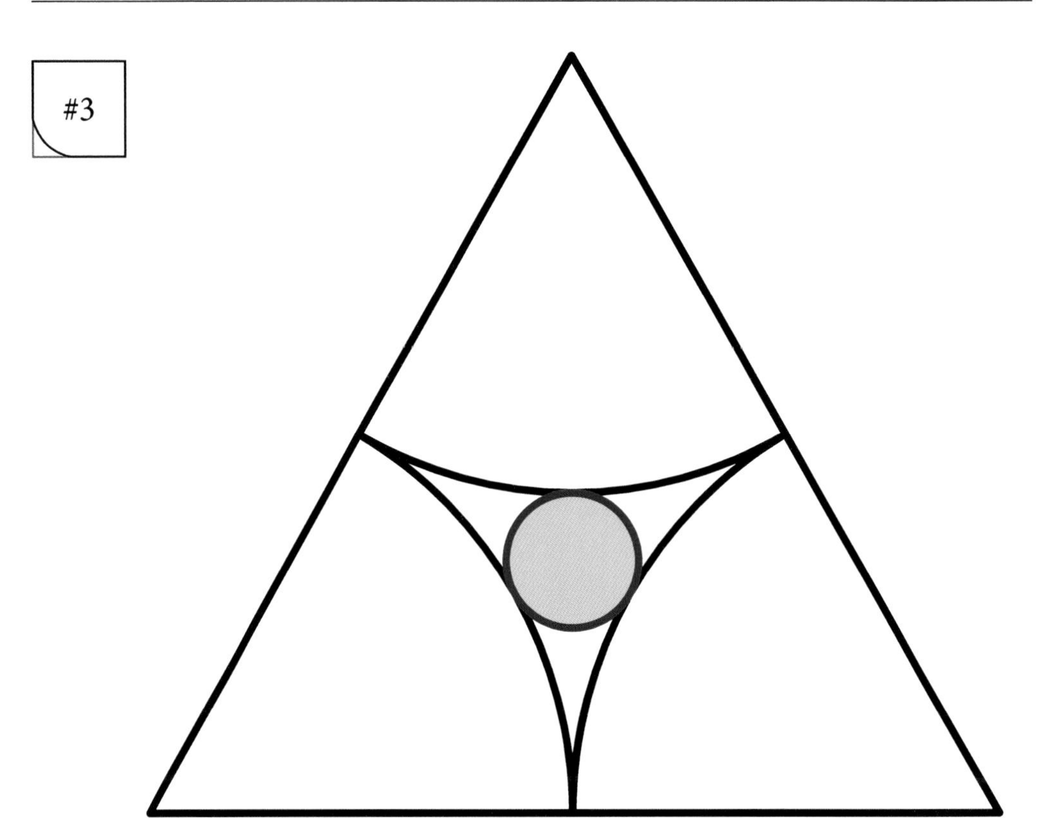

A circle is constructed within an equilateral triangle of side length 2 using three congruent sectors as shown.

What is the radius of the circle?

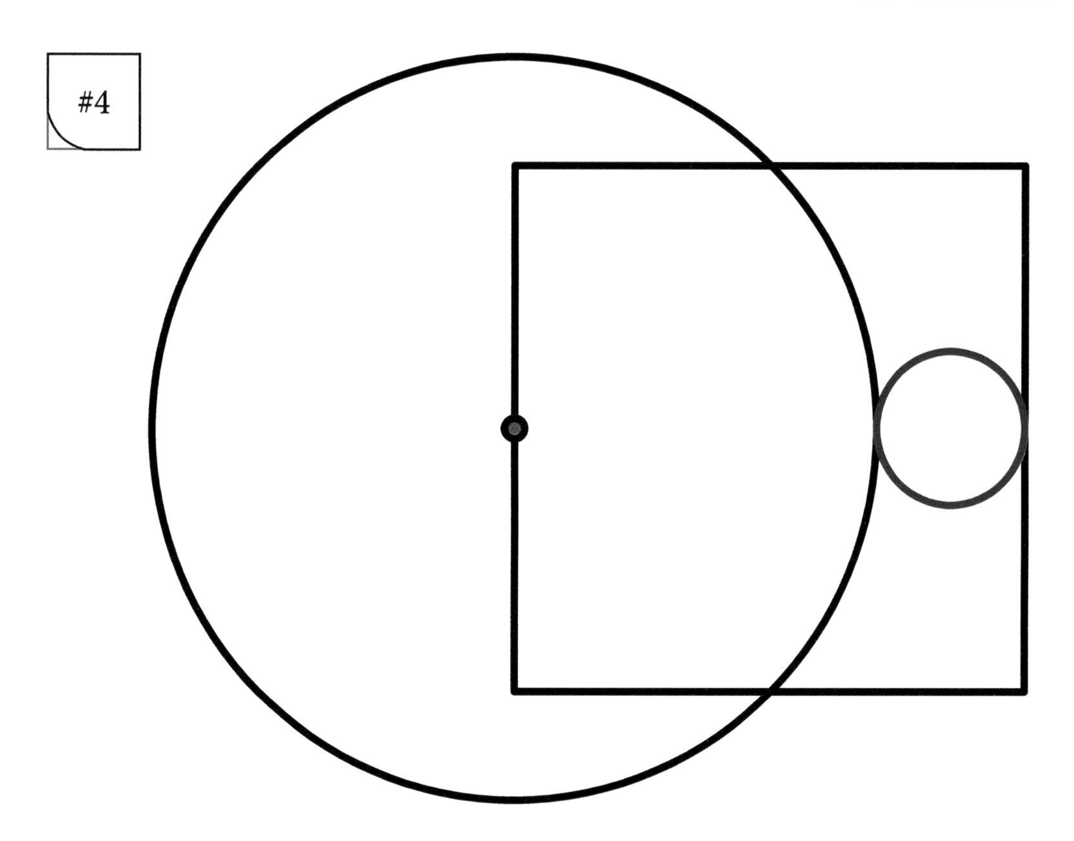

A circle is constructed using three midpoints of a square as shown. A unit circle is tangent to both the larger circle and the square.

Find the radius of the larger circle.

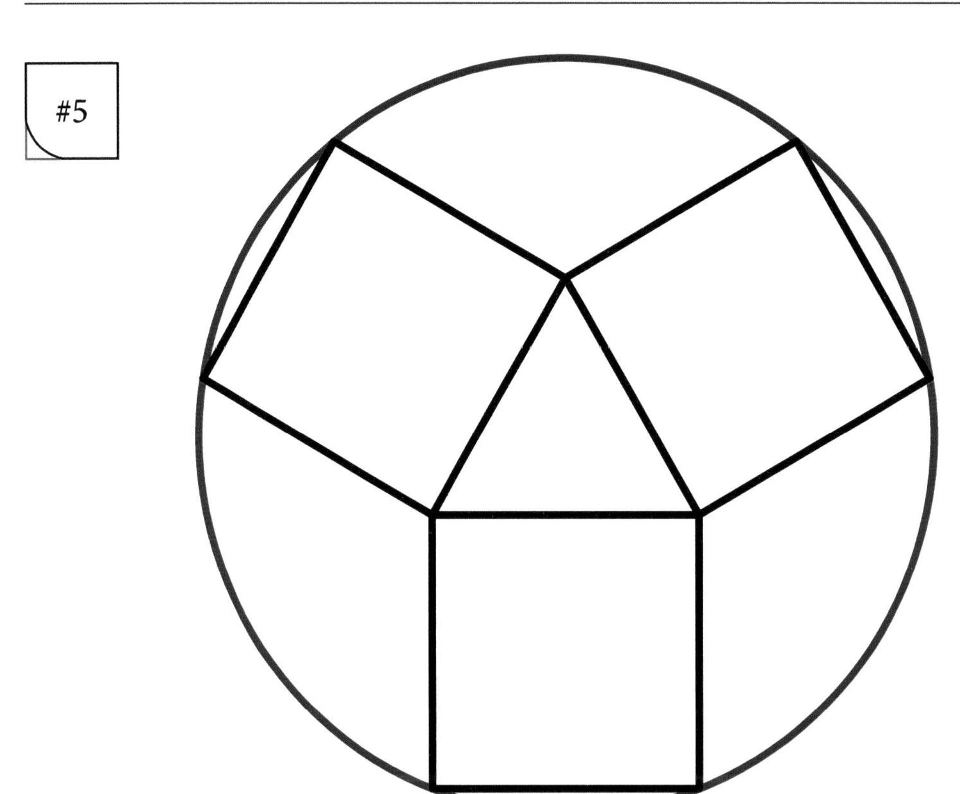

The sides of an equilateral unit triangle are used to construct three squares with vertices that touch a circle as shown.

Find the area of the circle.

#6

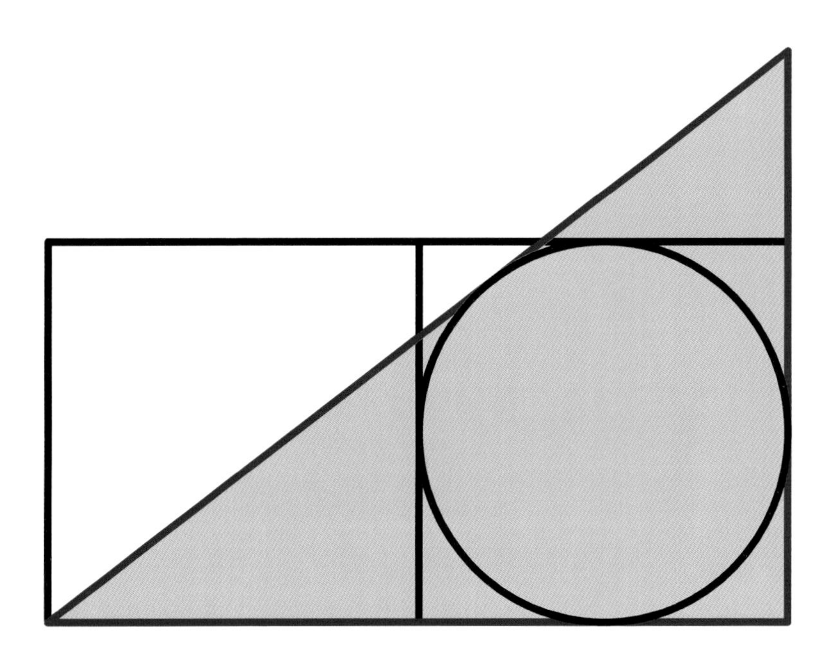

Two squares share a side, one circumscribing a unit circle to which the hypotenuse of the right triangle is tangent.

What is the area of the shaded right triangle ?

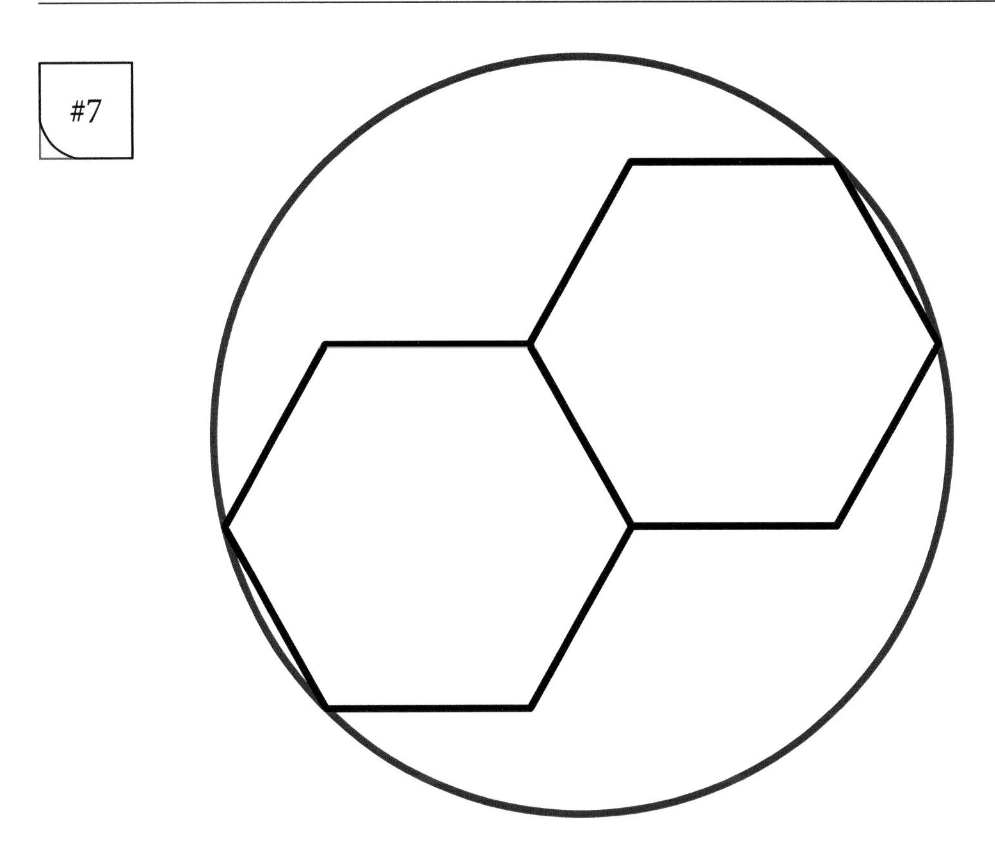

Two congruent regular unit hexagons are circumscribed as shown.

Find the radius of the circle.

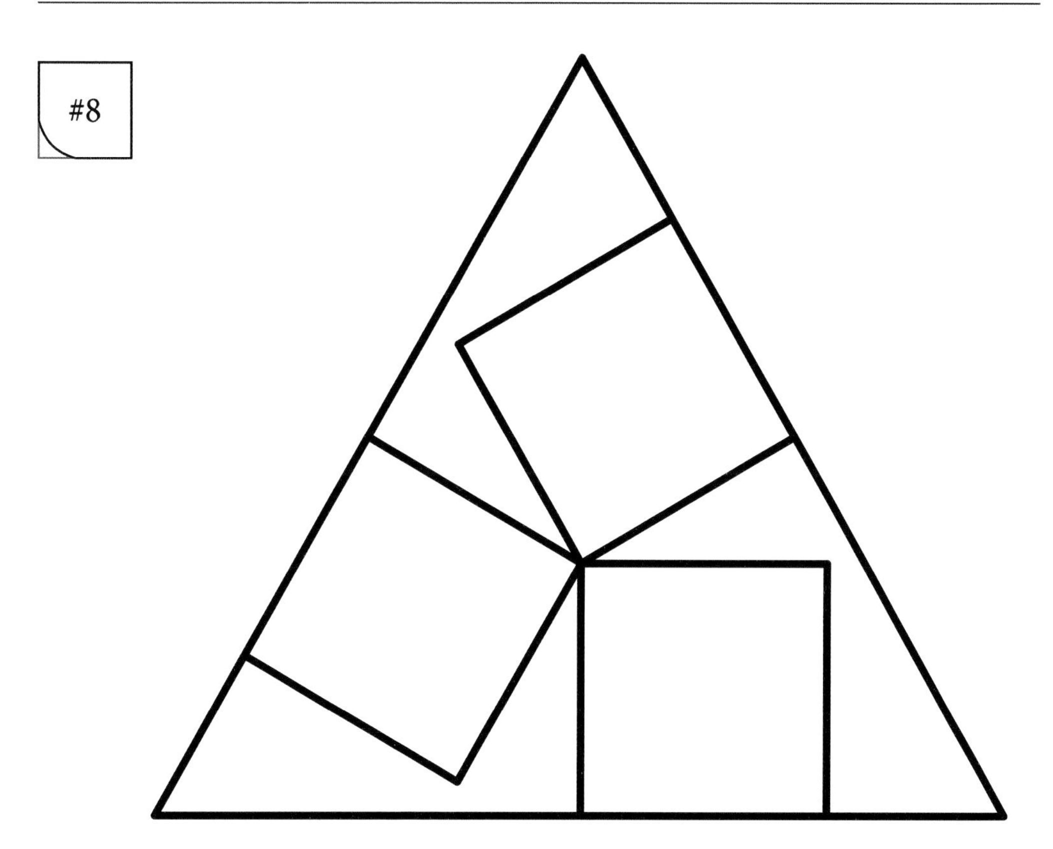

Unit squares are constructed on each side of an equilateral triangle such that they share a single vertex as shown.

Find the area of the triangle.

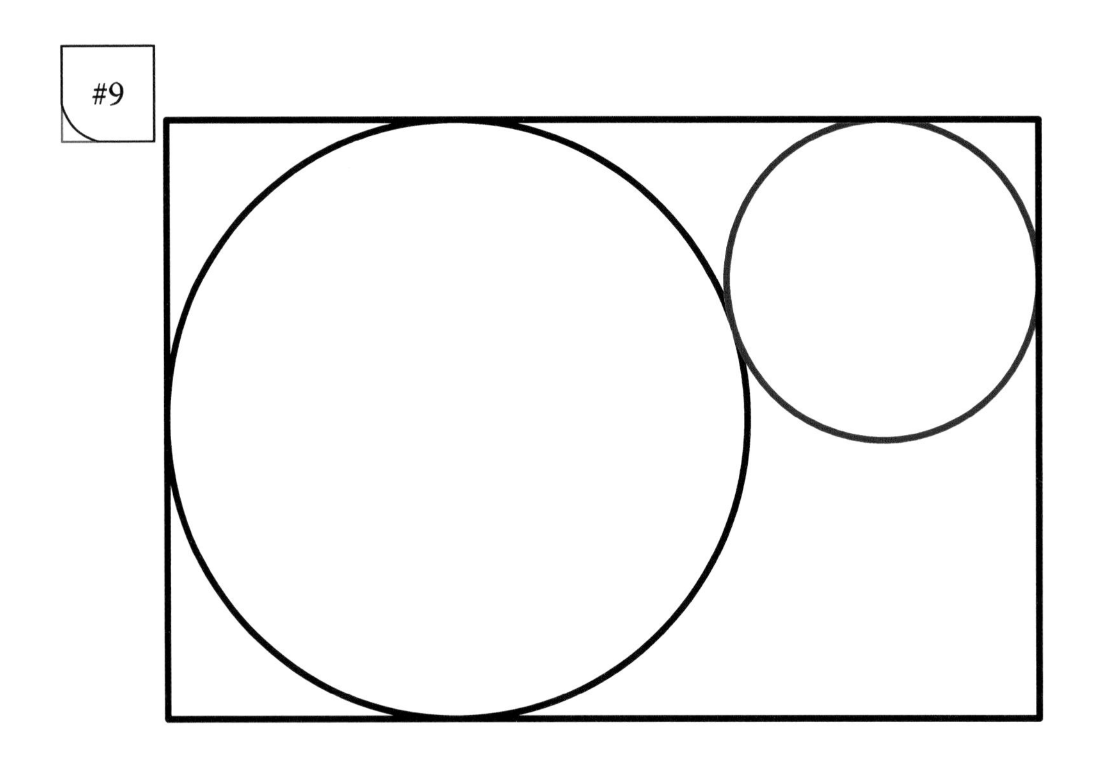

Cotangent circles are enclosed in a rectangle as shown.

If the radius of the large circle is 3, and the length of the rectangle is 8, what is the radius of the small circle?

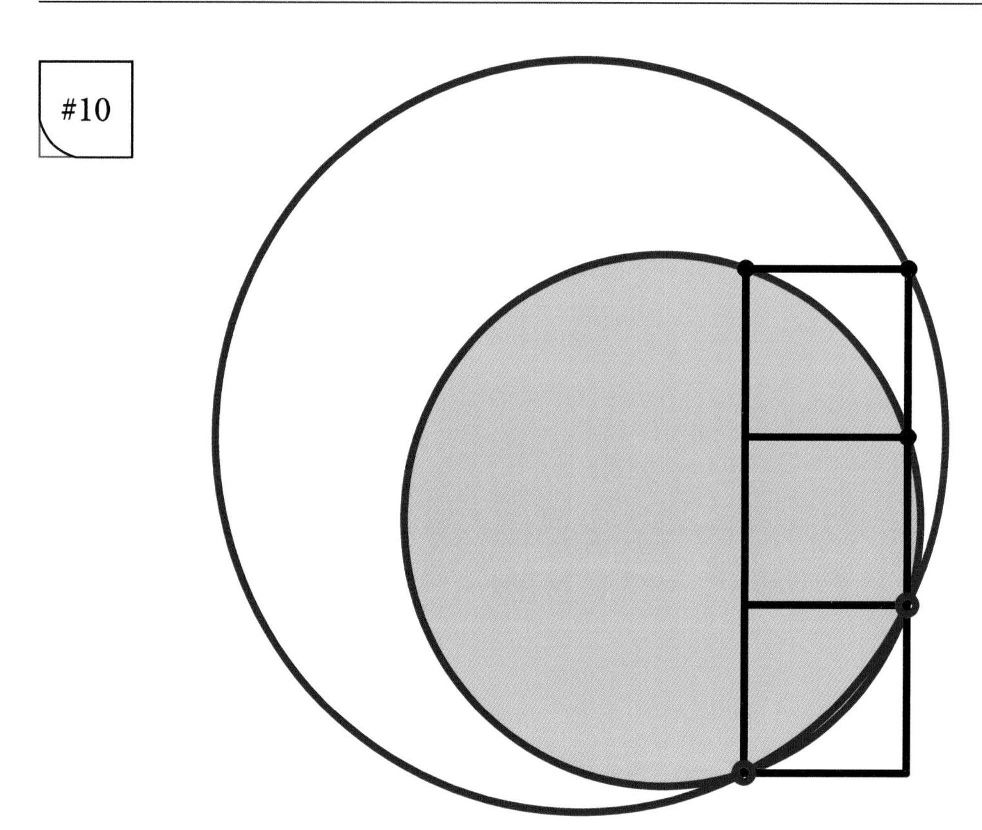

Three unit squares are stacked, sharing a side. Two circles each go through the marked vertices as shown. Both pass through the two vertices of the bottom square.

Prove that the area of the big disk is twice the area of the shaded one.

Solutions and Answers

#1. Answer: 1

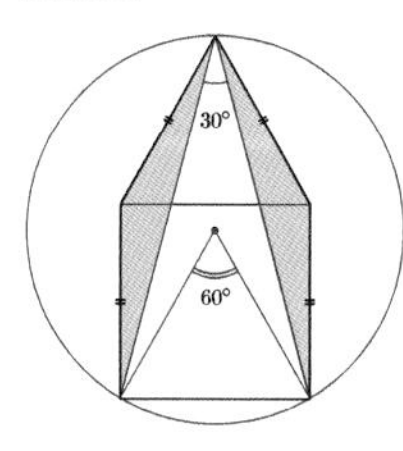

The two highlighted isosceles triangles have base angles of $(180 - 90 - 60) \div 2 = 15°$. This explains the 30° marked angle. Therefore the angle from the centre of the circle to the vertices at the bottom of the square must be 60° and thus, a second equilateral triangle is formed, with radius equal to the sides of the square.

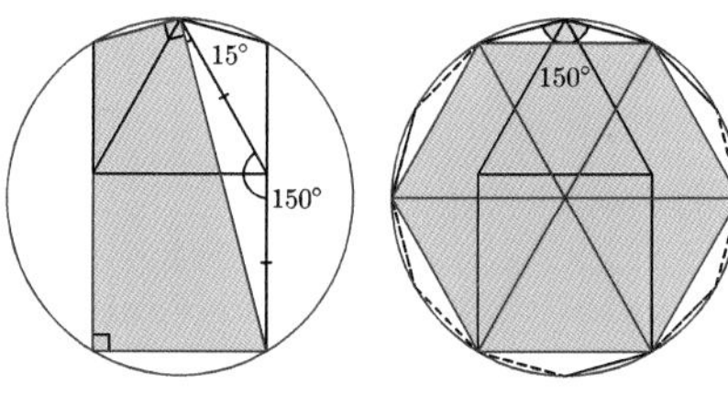

Using the cyclic quadrilateral shown, we can derive that the top right angle is 90°. Combining this with the leftover angle within the triangle (which is 15°) we can determine that the entire top angle is 150°. We can therefore construct a circumscribed dodecagon, whose alternate vertices form a regular hexagon, which is itself constructed from six equilateral triangles. The centre of these is the centre of the circle, and they are of length 1 as indicated by the square at the bottom of the diagram.

#2. Answer: $4 + \frac{7\sqrt{3}}{3}$

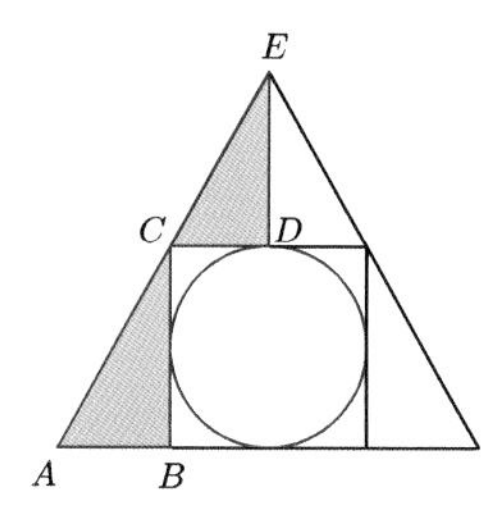

The triangle ABC has angles 60°, 90° and 30°, and therefore side ratios of $1 : \sqrt{3} : 2$. We know BC is length 2, therefore AB must be $\frac{2\sqrt{3}}{3}$. From here we can determine the base of the triangle as $2 + 2(\frac{2\sqrt{3}}{3})$. Using the area formula for an equilateral triangle $\frac{a^2\sqrt{3}}{4}$ yields $(4 + \frac{7\sqrt{3}}{3})$.

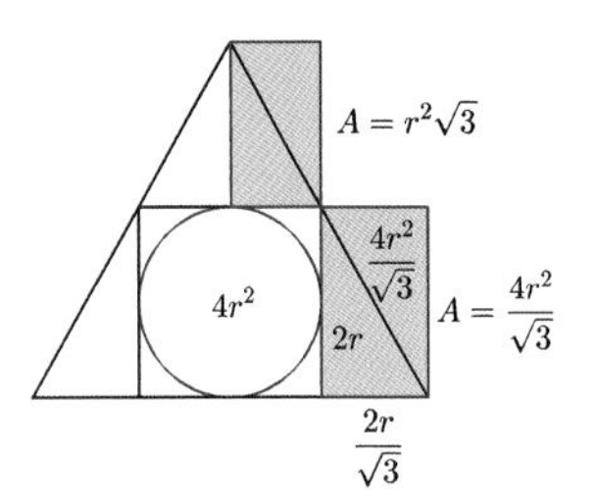

The area of the square is 4. The area of the rectangle formed by rotating one half of the triangle above the square is $\frac{2^2\sqrt{3}}{4} = \sqrt{3}$ and the area of the bottom two remaining triangles combined is $\frac{4}{\sqrt{3}}$.

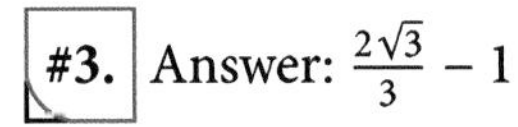

#3. Answer: $\frac{2\sqrt{3}}{3} - 1$

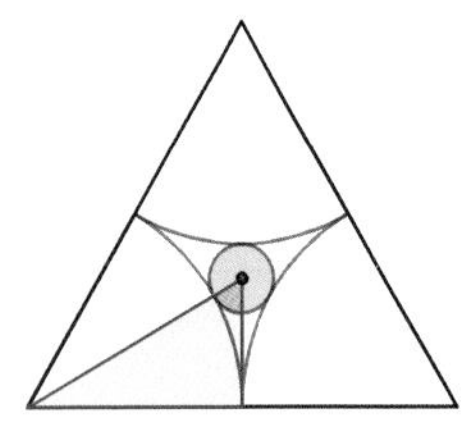

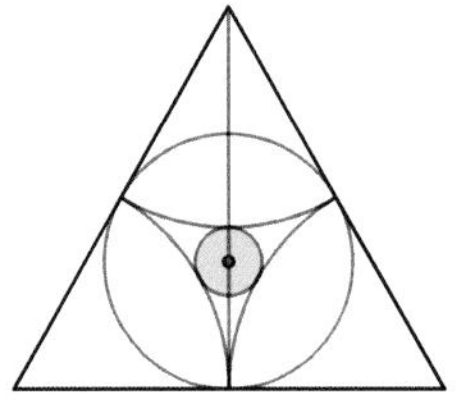

Using the 30, 60, 90 triangle indicated, we know the base is 1 and therefore can create an equation for the hypotenuse using the ratio highlighted in the previous solution, which is $1 + r = \frac{2\sqrt{3}}{3}$ which rearranges to the answer.

The circle and the incircle are concentric. Therefore their centre lies $\frac{2}{3}$ down the line shown (the height). So the length from the apex to the centre of the circle can be written as $(\frac{2}{3})(\frac{2\sqrt{3}}{2}) = 1 + r$. Solving this yields $r = \frac{2\sqrt{3}}{3} - 1$.

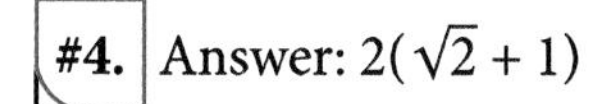

#4. Answer: $2(\sqrt{2} + 1)$

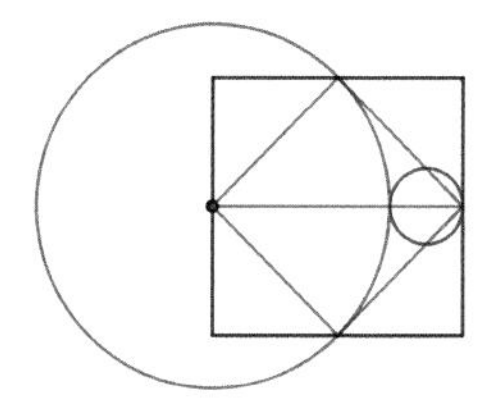

Assume instead that the radius of the larger circle is 1. Then we can demonstrate that the diameter of the smaller circle would be $\sqrt{2} - 1$ using Pythagoras and the isosceles right triangle shown. Therefore the ratio of the two radii would be $1 : \frac{\sqrt{2}-1}{2}$. This ratio must always be true, so if the radius of the smaller circle is 1, then the radius of the bigger circle is $\frac{2}{\sqrt{2}-1} = 2(\sqrt{2} + 1)$.

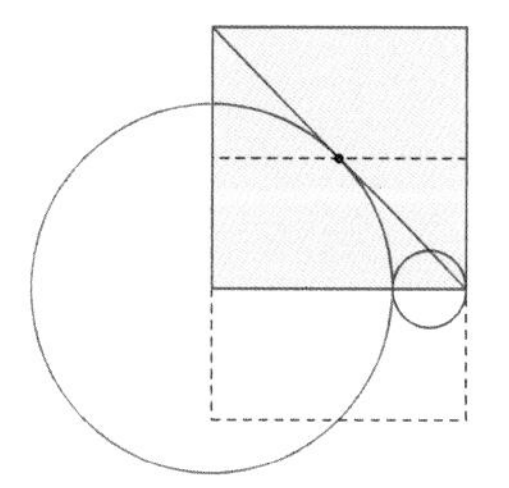

By reflecting half of the square as shown, we can see that the side lengths are $2 + r$ and the hypotenuse of the right triangle formed is $2r$. Use Pythagoras and solve the resulting equation $r^2 - 4r - 4 = 0$ to yield the answer.

#5. Answer: $\frac{\pi}{3}(4 + \sqrt{3})$

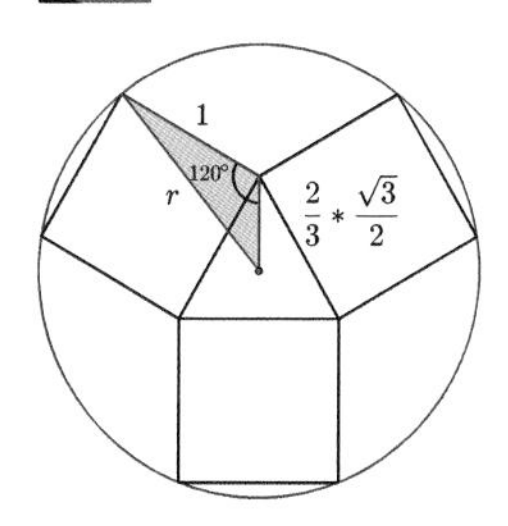

The centres of the circle and the triangle are the same, and so we can form a triangle with side lengths r, 1, $(\frac{2}{3})(\frac{\sqrt{3}}{2})$ and we can identify one angle as 120°. Using the cosine rule, we can determine that $r^2 = 1^2 + (\frac{\sqrt{3}}{3})^2 + \frac{\sqrt{3}}{3})$. Multiply this by Pi and the answer is $\frac{\pi}{3}(4 + \sqrt{3})$.

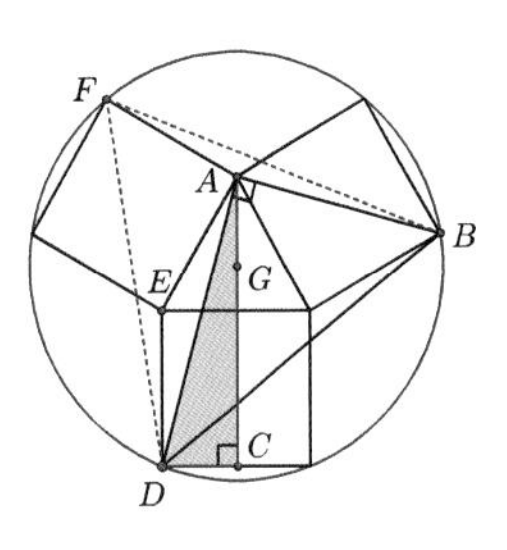

Calculate length DA using $\sqrt{(\frac{1}{2})^2 + (1 + \frac{\sqrt{3}}{2})^2} = \frac{\sqrt{2}}{2}(1 + \sqrt{3})$. Note that angle $DAB = 90°$ (you can derive this using $CAB = 45 + 30$ and $DAC = EAD = \frac{180-150}{2}$. Therefore we can derive DB (using Pythagoras) as $\sqrt{4 + \sqrt{3}}$. This is the side length of the equilateral triangle DFB whose centre is the centre of the circle. Hence the radius is $\frac{2}{3}$ the height of the triangle. Further calculations are left to the reader or can be found online (https://www.tarquingroup.com/geomsnackssolution/).

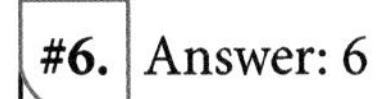

#6. Answer: 6

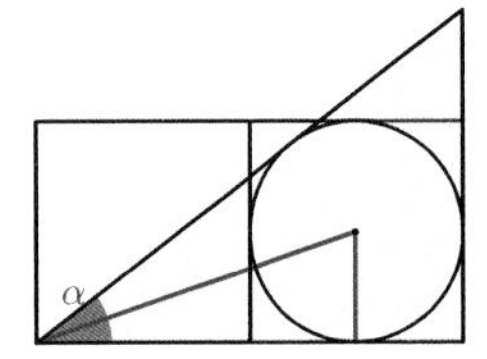

By drawing two radii and a bisector, one sees that: $\tan\left(\frac{\alpha}{2}\right) = \frac{1}{3}$. Using the formula $\tan(\alpha) = \frac{2\tan(\alpha/2)}{1-\tan^2(\alpha/2)}$ we find: $\tan(\alpha) = \frac{3}{4}$. The adjacent side of the shaded triangle having a length of 4, the opposite side has a length of 3 and the area is therefore 6.

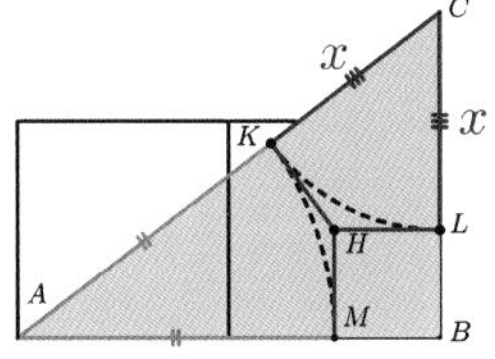

Choosing $x = CK$, notice that $CK = CL = x$ and $AK = AM = 3$. The Pythagorean theorem gives you $(x + 3)^2 = 4^2 + (x + 1)^2$, the x^2 cancel out to give you $6x + 9 = 16 + 2x + 1$, so $x = 2$. The shaded triangle has sides measuring 3, 4, 5. Its area is therefore 6.

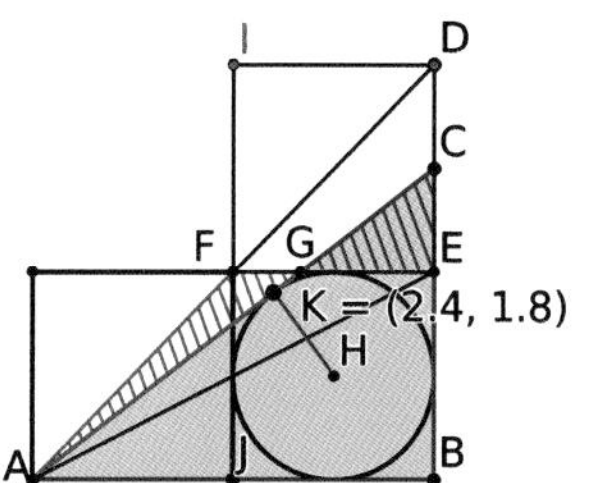

Using a natural coordinate system (origin A, $AJ = 2$), draw a third square $EFID$ and consider C the midpoint of the side FD. Claim: (AC) is tangent to the circle in $K(2.4, 1.8)$, i.e. $K(3 - \frac{3}{5}, 1 + \frac{4}{5})$. Indeed: $\overrightarrow{HK}(-\frac{3}{5}, \frac{4}{5})$ so $HK = 1$ and the slope of (HK) is $-\frac{4}{3}$. Since (AC) has equation $y = \frac{3}{4}x$: $K \in (AC)$ and $(HK) \perp (AC)$. So G is the centroid of ΔAED, hence $GE = 2FG$. The hatched triangles have the same area because one base is half the other's while one altitude is twice the other's. The area of ΔABC is therefore the same as that of $ABEF$ ie a square and a half: 6.

#7. Answer: $\frac{\sqrt{13}}{2}$

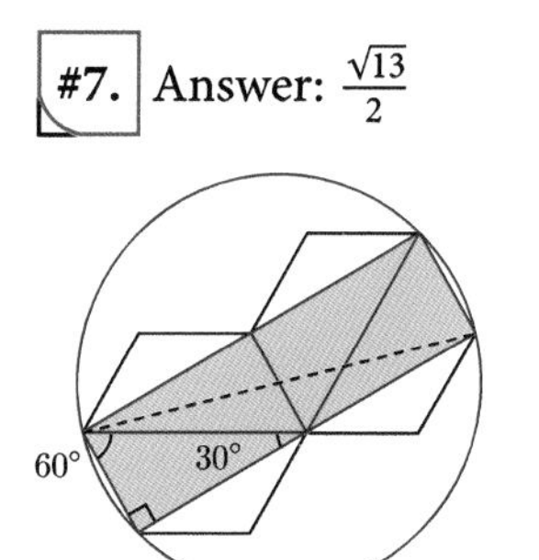

Construct a rectangle made up of four congruent right triangles as shown. We know the shorter side of the triangles is 1 and as they're 30°, 60°, 90° triangles, then their side ratios are $1 : 2 : \sqrt{3}$. This means the rectangle must have length $2\sqrt{3}$ and width 1. Therefore, using Pythagoras, the diagonal of the rectangle (the diameter of the circle) is $\sqrt{13}$.

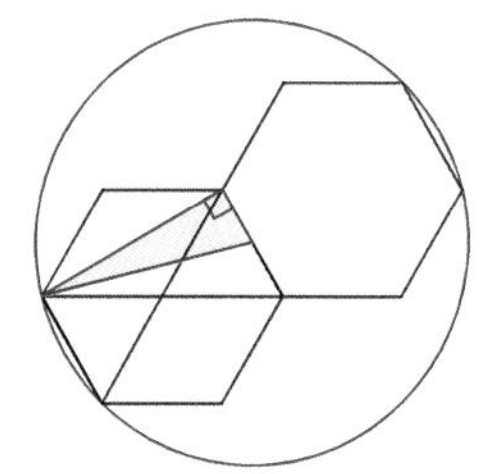

Using only one of the hexagons, we can construct a right triangle as shown, which has the height of two equilateral unit triangles ($\sqrt{3}$) and base of $\frac{1}{2}$. Using Pythagoras we can determine the hypotenuse (radius) as $\frac{\sqrt{13}}{\sqrt{4}}$.

#8. Answer: $3\sqrt{3}$

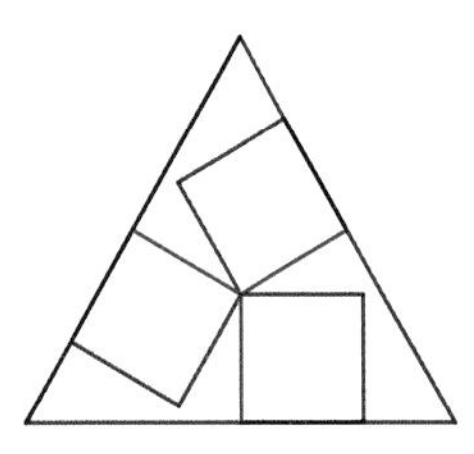

Using Viviani's Theorem (highlighted in the original Geometry Snacks), the shortest lengths from each side to an internal point (for this problem, where the squares meet) sum to the height of the triangle. As the sides of the square are perpendicular to the sides of the triangle, then these are the shortest lengths. Hence the height of the triangle must be 3 which gives us the equation $\frac{a\sqrt{3}}{2} = 3$. So $a = 2\sqrt{3}$ and the area must be $3\sqrt{3}$.

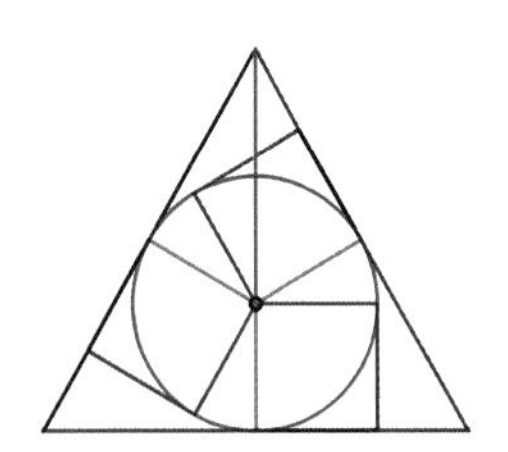

Constructing the incircle of the triangle enables us to see that its diameter is 2, and therefore the height of the triangle is 3. From here solve as with the last approach.

#9. Answer: $11 - 4\sqrt{6}$

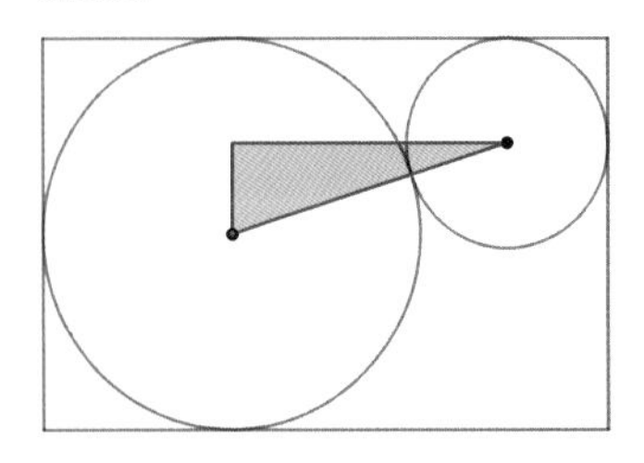

By constructing a right triangle as shown, we can deduce that the side lengths are $3 - r$, $5 - r$ and $3 + r$. From this we can form a quadratic equation using Pythagoras: $r^2 - 22r + 25 = 0$. Solving this using the quadratic formula gives $11 \pm 4\sqrt{6}$. Since r is less than 6, then $r = 11 - 4\sqrt{6}$.

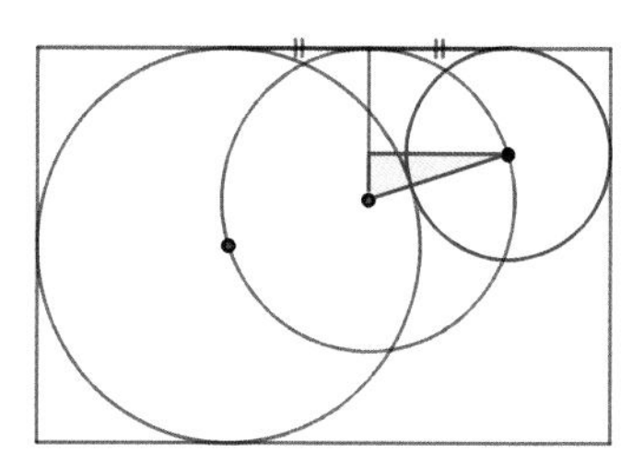

Constructing a new circle using the midpoint between the centres of each original circle, we can construct a similar triangle as before, but this time the sides are half the values in our first solution. From here solve in much the same way.

#10. Answer:

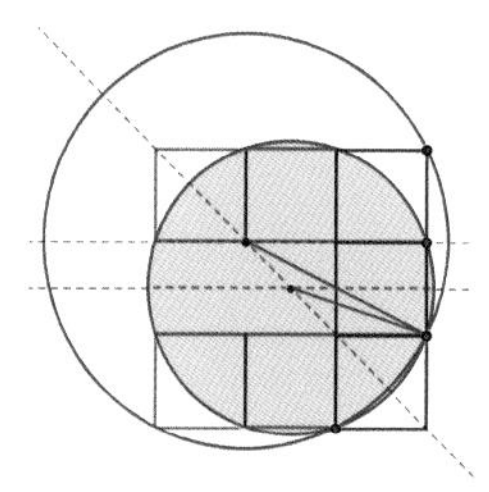

Drawing the perpendicular bisectors of three chords, one can find the position of the centres and using them as lines of reflection gives the squares shown. We can now find the radii of both circles using the Pythagorean theorem. $R^2 = 2^2 + 1^2 = 5$ and $r^2 = (\frac{3}{2})^2 + (\frac{1}{2})^2 = \frac{5}{2}$. So $R^2 = 2r^2$ which gives the sought result.

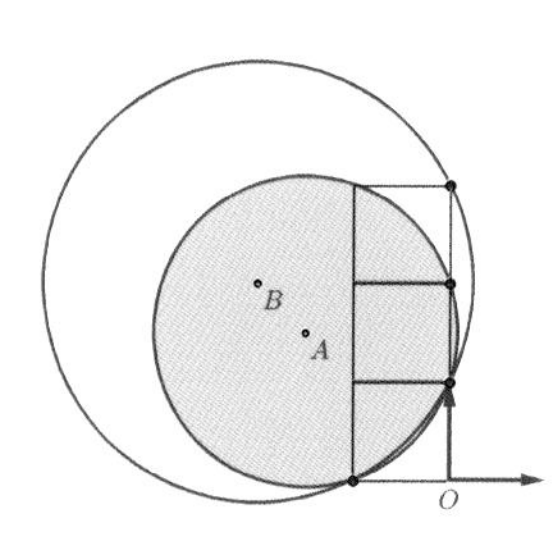

Using a coordinate system with origin O as shown you can write these equations of the two circles with centres $A(x_A, y_A)$ and $B(x_B, y_B)$:

$$(x - x_A)^2 + (y - y_A)^2 = r^2 \quad \text{and} \quad (x - x_B)^2 + (y - y_B)^2 = R^2.$$

Using the points $(-1, 0)$ and $(0, 1)$ common to the two circles:

$$(-1 - x_A)^2 + y_A^2 = x_A^2 + (1 - y_A)^2.$$

Expanding this you get $y_A = -x_A$ and $y_B = -x_B$. Using $(0, 2)$ and $(0, 1)$:

$$x_A^2 + (2 - y_A)^2 = x_A^2 + (1 - y_A)^2$$

so $2 - y_A = y_A - 1$, i.e. $y_A = \frac{3}{2}$. Similarily, using $(0, 3)$ and $(0, 1)$ gives $y_B = 2$. Finally $A(-\frac{3}{2}, \frac{3}{2})$ and $B(-2, 2)$; plugging these coordinates in the equations you get: $r^2 = \frac{1}{4} + \frac{9}{4} = \frac{5}{2}$, $R^2 = 5$. So $R^2 = 2r^2$ which solves the problem.

3D Geometry

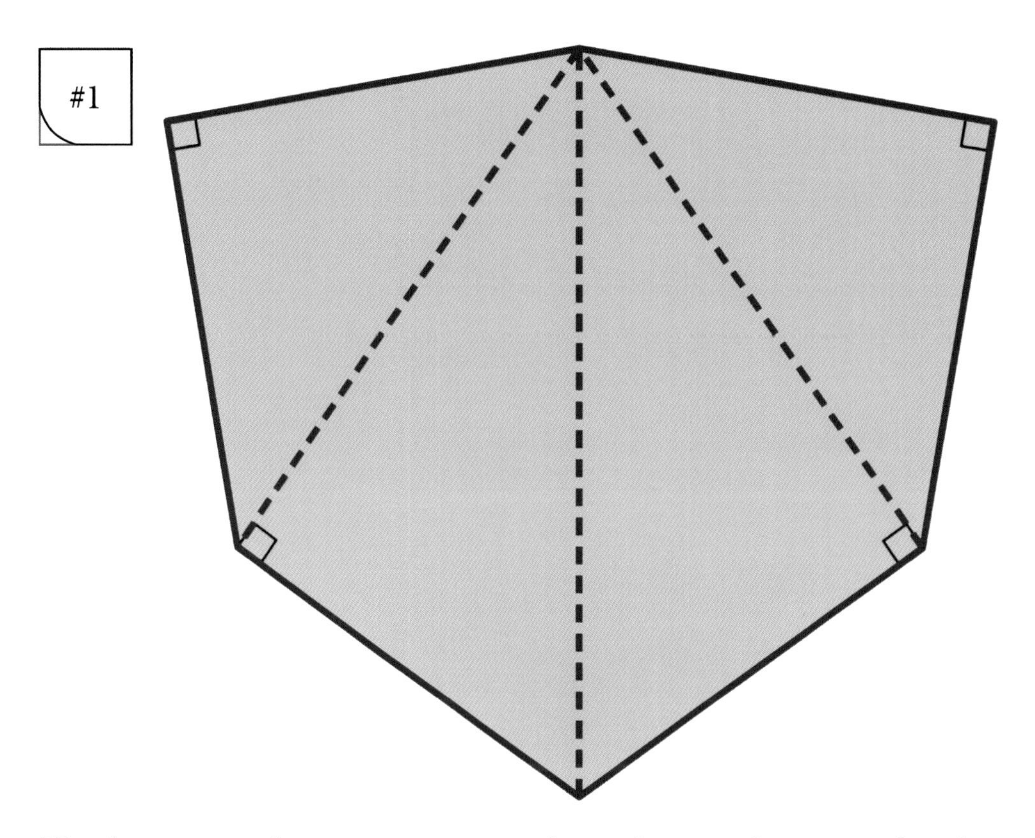

This hexagon with congruent unit sides is the net of a pyramid with a square base (not drawn).

What is the volume of this pyramid?

#2

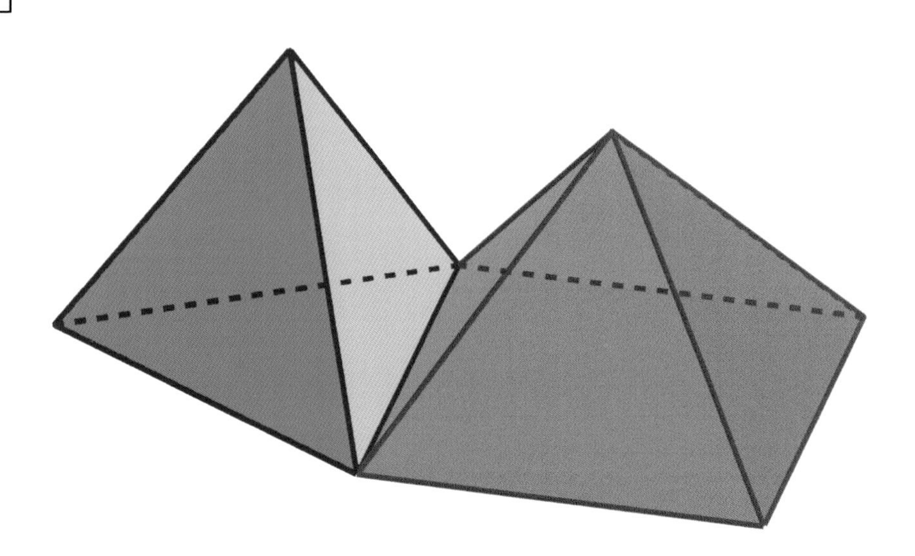

A regular tetrahedron and a square based pyramid are shown, all the edges have unit length.

Express the volume of one as a fraction of the other.

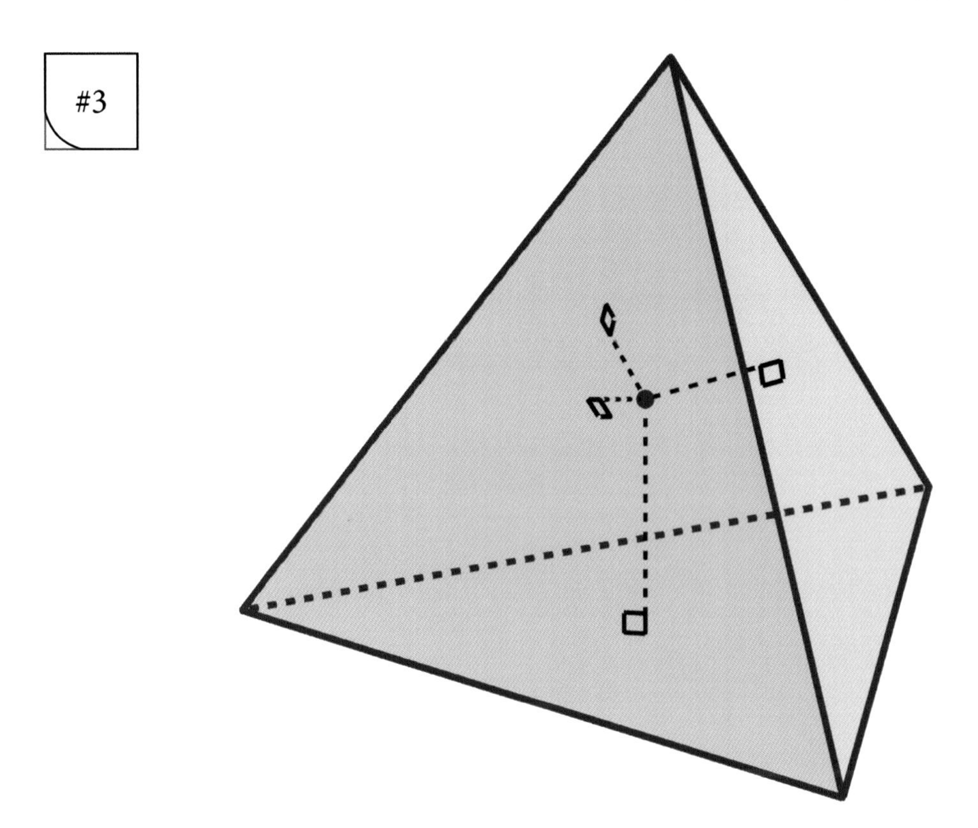

Prove that from an interior point of a regular tetrahedron, the sum of distances to the four faces does not depend on the chosen point.

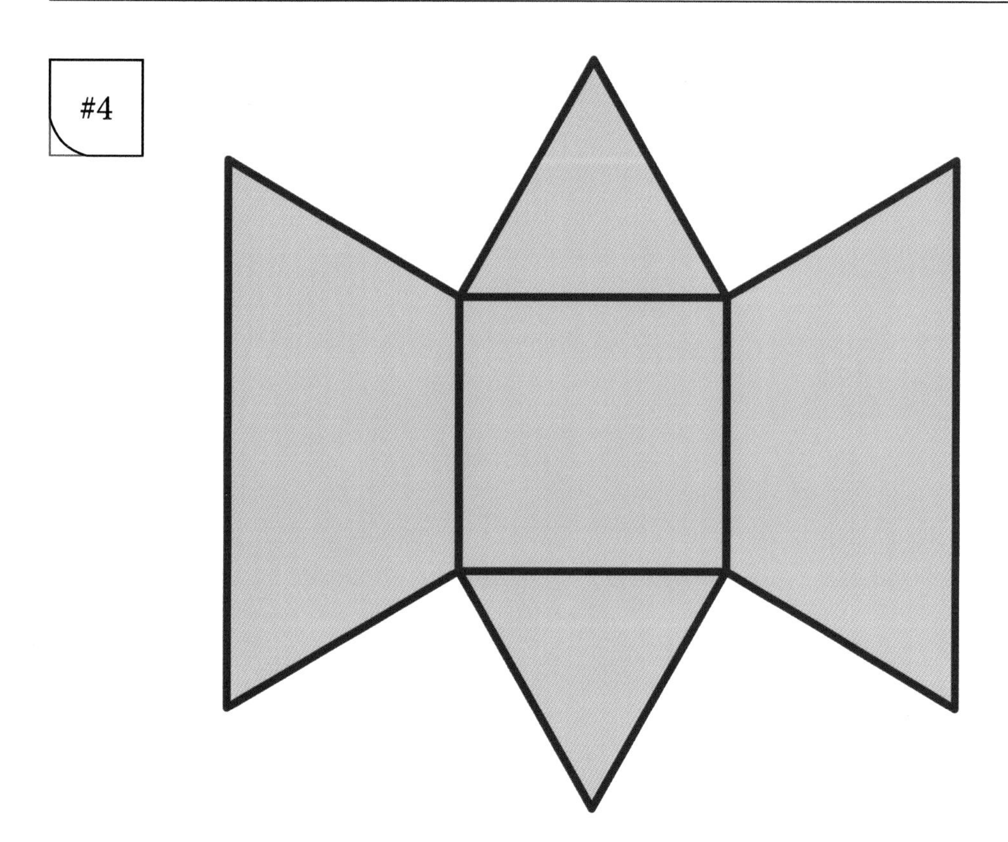

What is the volume of the polyhedron that has such a net?

The central shape is a unit square with equilateral triangles and trapeziums around it. The long edge is twice as long as the short ones.

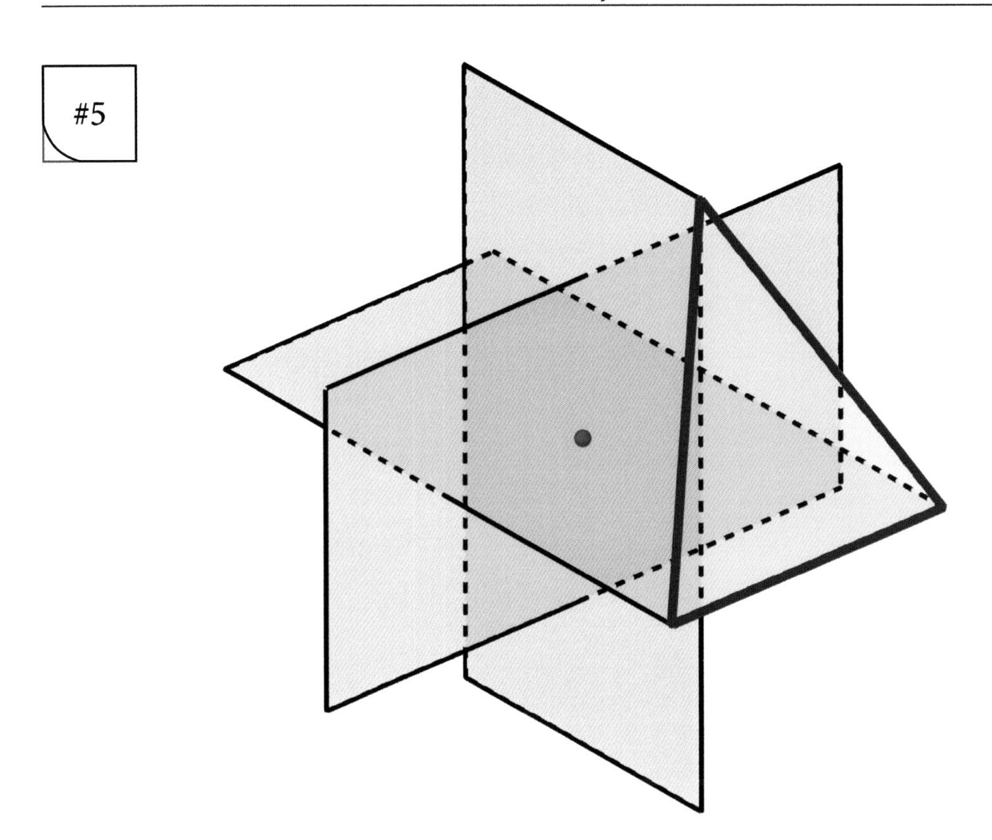

Three congruent rectangles have a common centre and are perpendicular to the other two as shown.

What should the ratio length:width of these rectangles be so that the 12 vertices form a regular icosahedron (i.e. 20 equilateral faces)?

#6

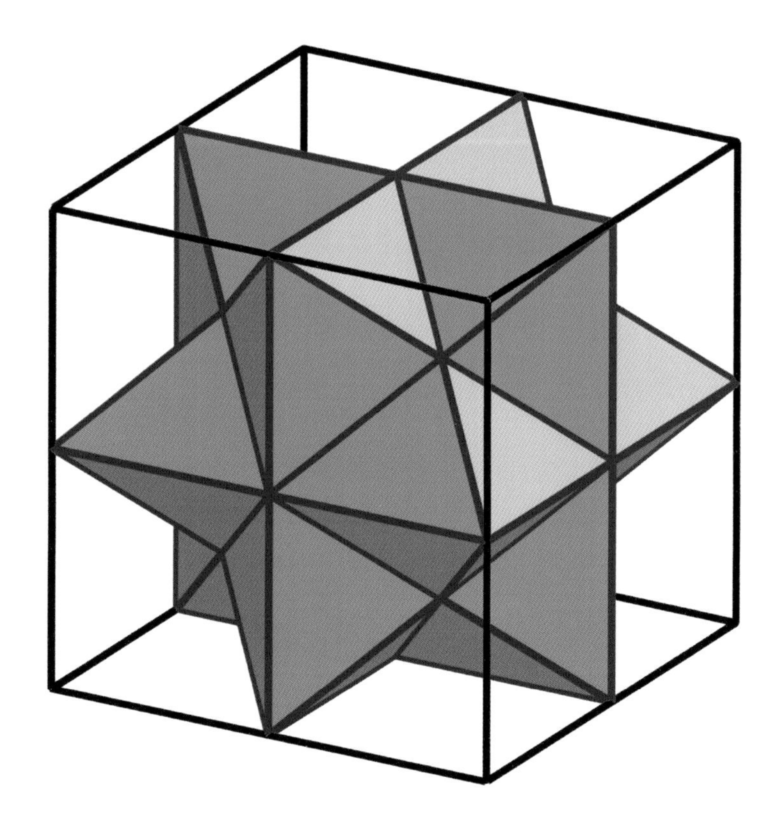

This star polyhedron is constructed inside a unit cube using the centre of the faces and the midpoints of the edges.

What is its volume?

#7

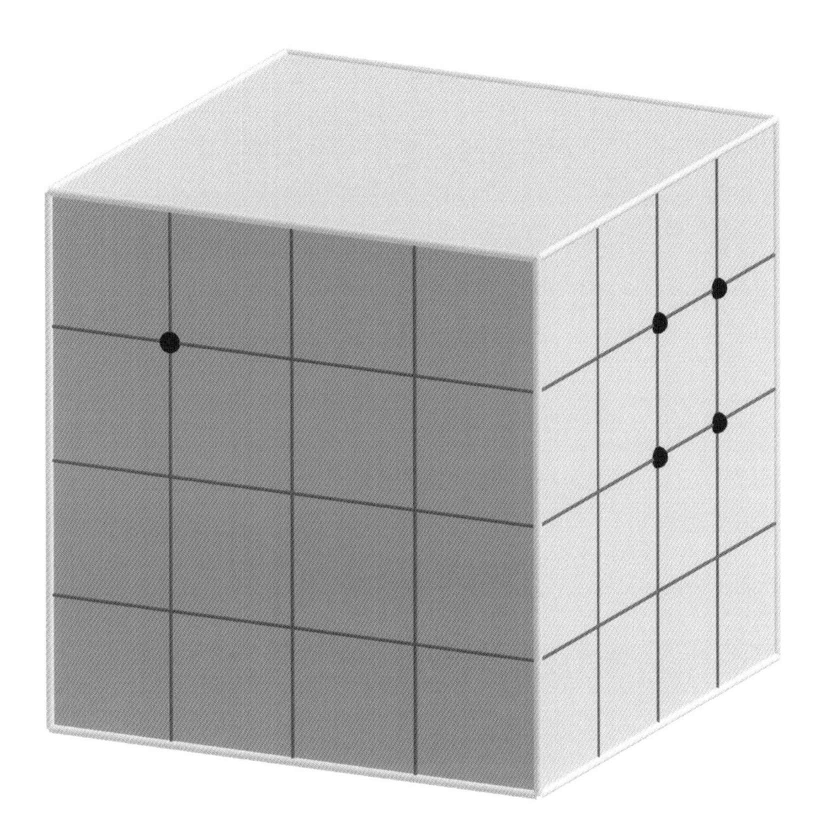

You want to connect the point on the front face of this cube to one of the four points on the right face by staying on the surface of the cube, whose edges measure 4 units.

For which of these four points will the shortest path require to go through the top face?

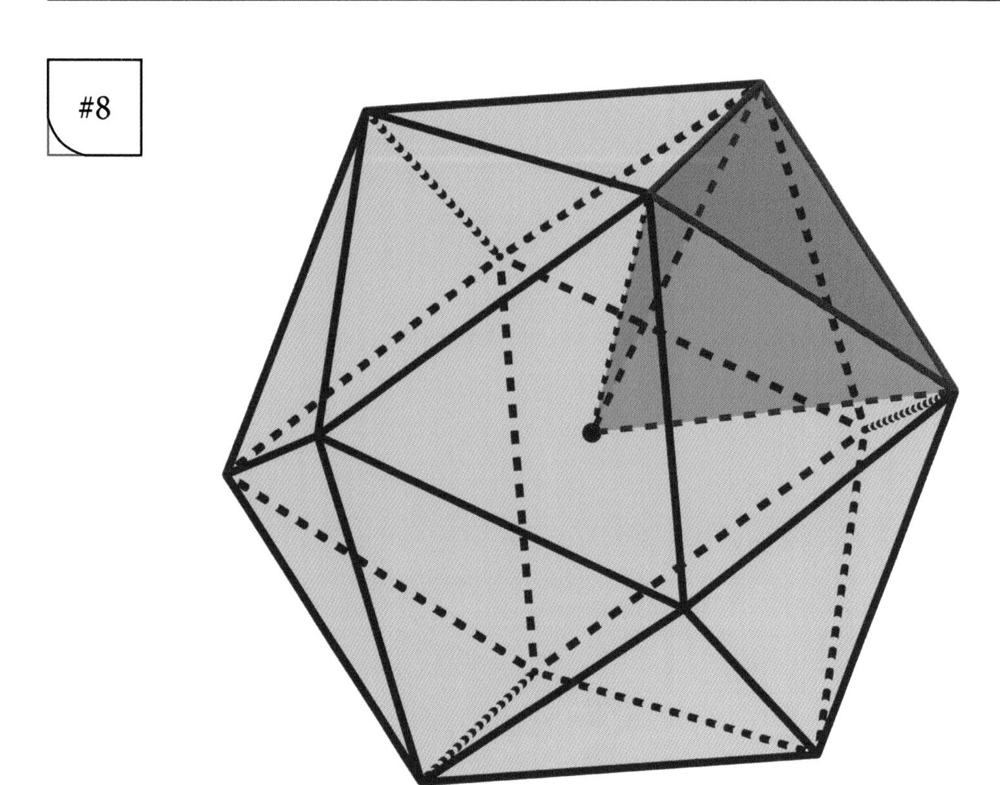

Is it possible to fill the regular icosahedron with 20 regular tertrahedra as the one shown, without any gaps or overlaps?

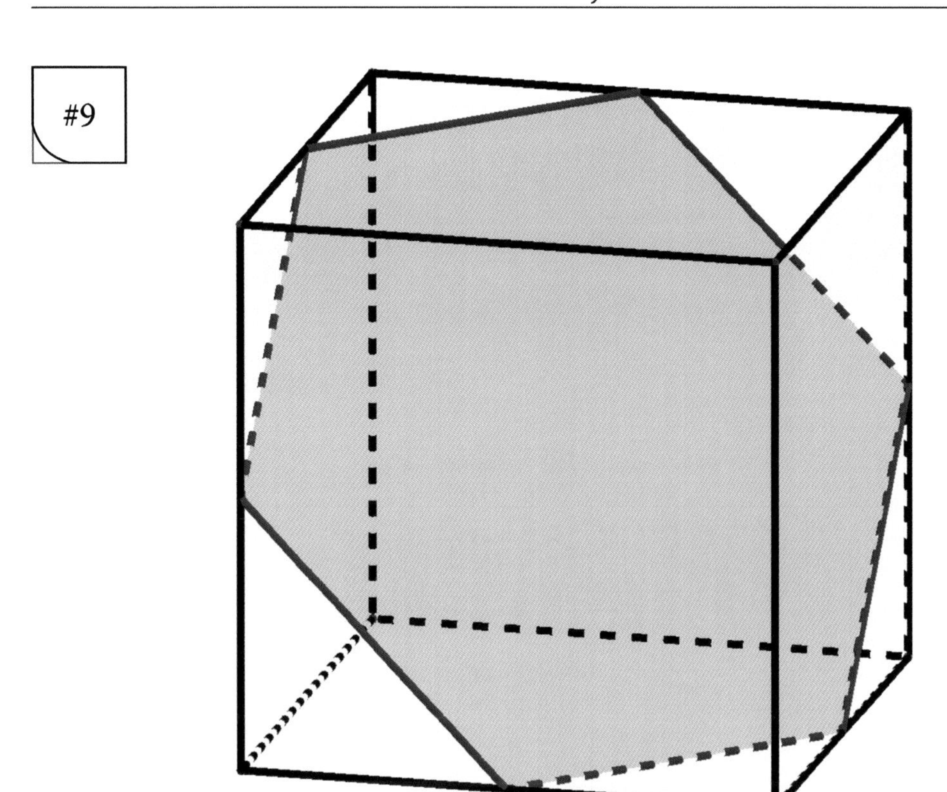

Can you find a net of this cube for which the regular hexagon that appears on its surface (going through midpoints of edges) is a straight line?

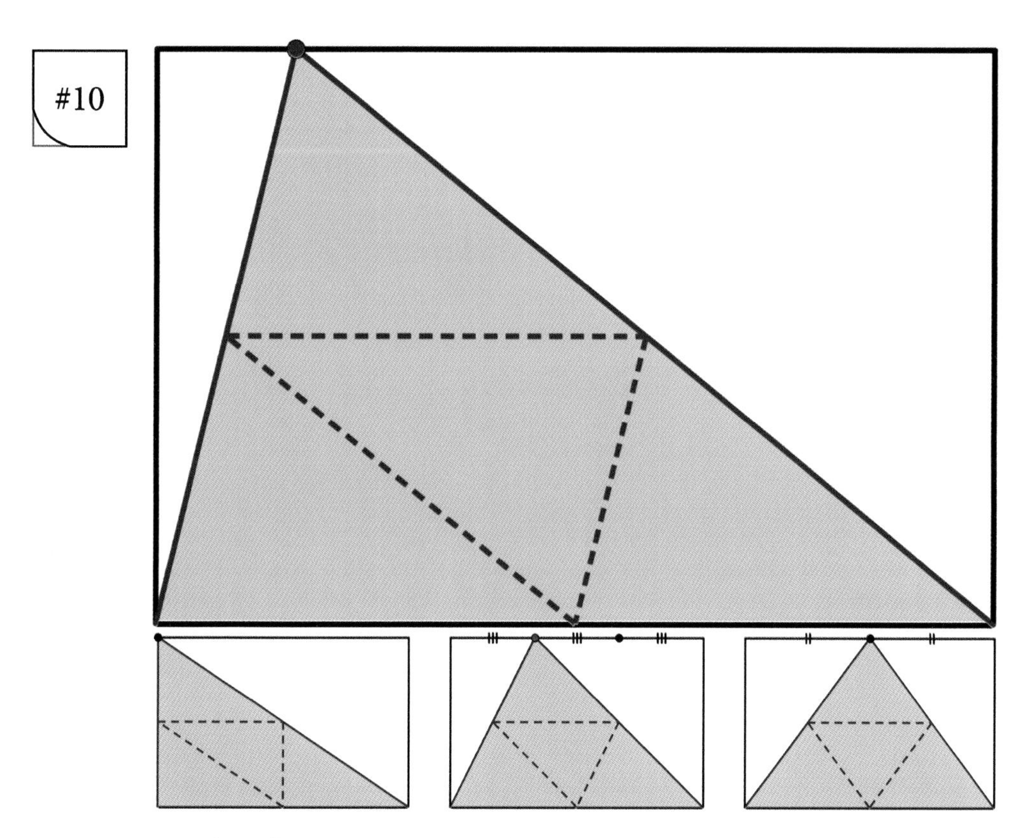

In a rectangular sheet 8×12 you choose a point on one long side and cut out the white corners. By folding the blue net using the midpoints, you want the greatest volume.

Among the three possibilities shown here, which one would you choose?

Solutions and Answers

#1. Answer: $\frac{1}{3}$

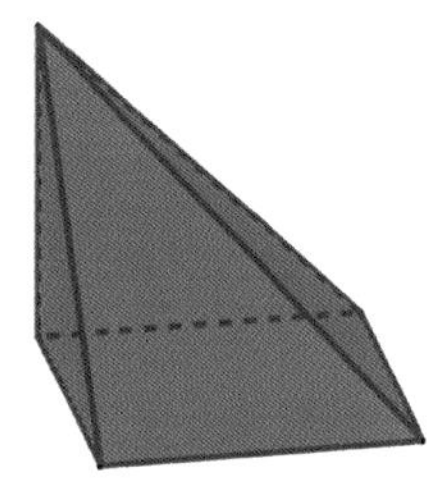

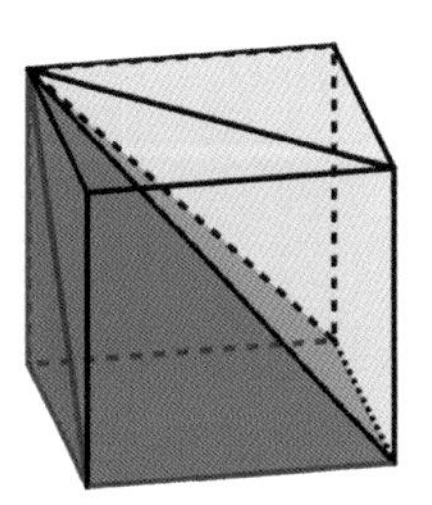

The two top sides of the net form the height of the pyramid because this line is perpendicular to two sides of the square base. The volume of a pyamid is one third of the area of the base multiplied by its height which gives here $\frac{1}{3}$.

Three such pyramids perfectly fit in a unit cube, so the volume of one is a third of the volume of the cube.

#2. Answer: $P = 2T$

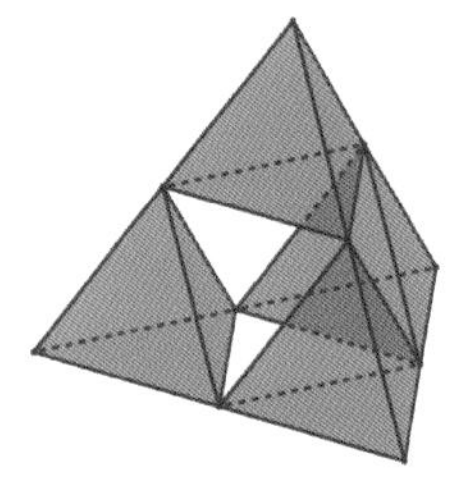

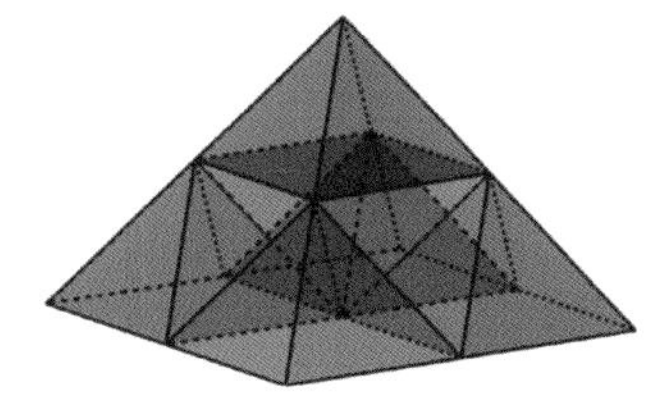

Four copies of the tetrahedron form a tetrahedron twice as big. The total volume is multiplied by $2^3 = 8$ but leaves a gap (which is an octahedron) that can be filled by two unit pyramids. So $8T = 4T + 2P$ gives $P = 2T$.

Same method with six pyramids leaving four gaps that can be filled by unit tetrahedra. $8P = 6P + 4T$ gives $P = 2T$. The authors thank Christian Mercat for these clever solutions.

#3. Answer:

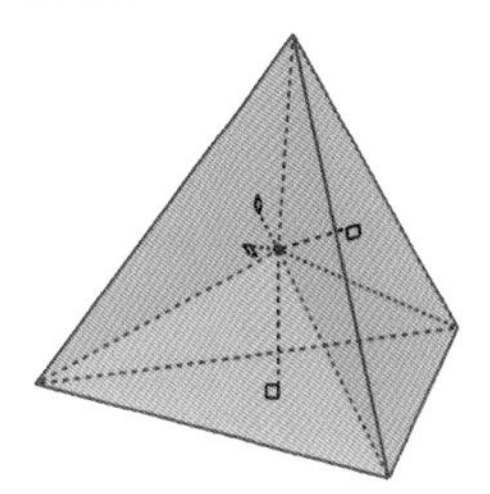

By connecting the interior point to the four vertices, the volume $\mathscr{T}$ of the tetrahedron is split in four tetrahedra $\mathscr{T}_1, \mathscr{T}_2, \mathscr{T}_3, \mathscr{T}_4$. Their volume can be written $\frac{1}{3}\mathscr{A} \times h_i$ where $\mathscr{A}$ is the area of an equilateral face. So:

$$\mathscr{T} = \frac{1}{3}\mathscr{A} \times (h_1 + h_2 + h_3 + h_4).$$

So $h_1 + h_2 + h_3 + h_4$ is equal to the height of the tetrahedron, whatever interior point we choose.

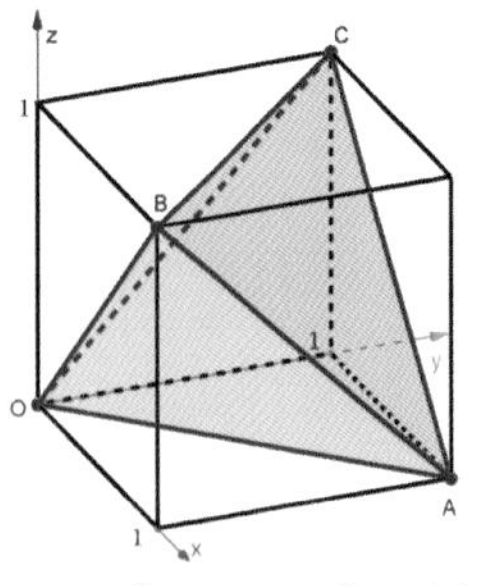

Let's consider the tetrahedron $OABC$ in this coordinate system with: $O(0,0,0)$, $A(1,1,0)$, $B(1,0,1)$ and $C(0,1,1)$. The points $M(x,y,z)$ inside the tetrahedron are the ones that satisfy:

$$x+y-z \geq 0; \quad x-y+z \geq 0; \quad -x+y+z \geq 0; \quad 2-x-y-z \geq 0.$$

These conditions determine the half-spaces containing $OABC$ with respective frontiers: (OBC), (OAC), (OAB), (ABC). The distance from a point $M(x,y,z)$ to a plane with equation $ax + by + c + d = 0$ is given by:

$$\frac{|ax + by + cz + d|}{\sqrt{a^2 + b^2 + c^2}}.$$

For the four planes' equations we have $\sqrt{a^2 + b^2 + c^2} = \sqrt{3}$ so the sum of distances is:

$$\frac{1}{\sqrt{3}}\left((x + y - z) + (x - y + z) + (-x + y + z) + (2 - x - y - z)\right) = \frac{2}{\sqrt{3}}.$$

This is independent from the point M chosen so it corresponds to the height of the tetrahedron.

Remark: This can be seen as a 3D version of Viviani's theorem. Notice that the proof still holds for any convex polyhedron with faces having the same area.

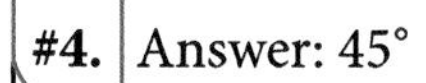

#4. Answer: 45°

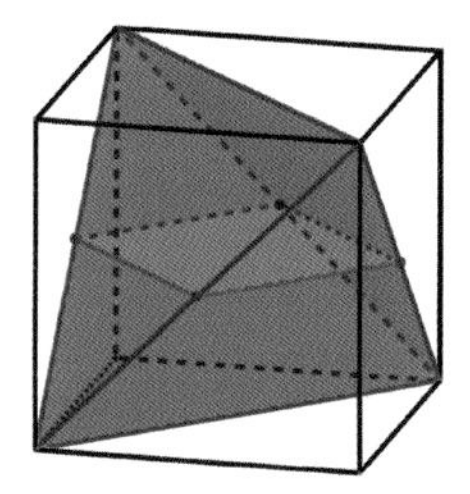

Two of the solids formed can be positioned to form a regular tetrahedron as shown inside this cube. If the cube has edge length ℓ, the tetrahedron's volume $\mathscr{T}$ can be deduced by subtracting four right pyramids.

$$\mathscr{T} = s^3 - 4 \times \frac{1}{3} \times \frac{s^2}{2} \times s = \frac{s^3}{3}$$

The sought volume is $\mathscr{V} = \frac{1}{2}\mathscr{T}$ for $s = \sqrt{2}$. $\mathscr{V} = \frac{2\sqrt{2}}{3}$.

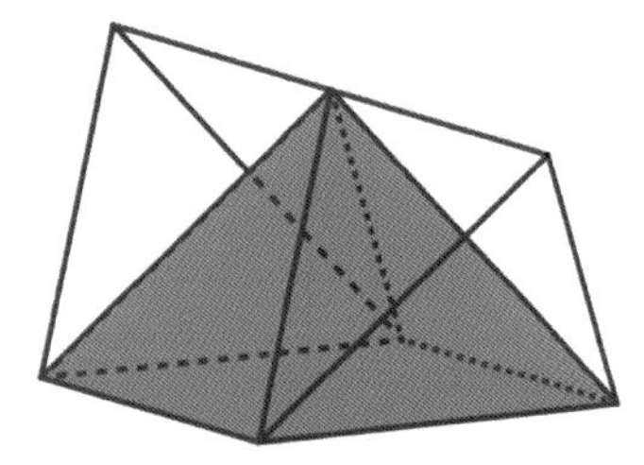

The solid can be seen as a regular square pyramid with two regular tetrahedra. We saw in the second 3D snack that this has the volume of two pyramids. The height is found via Pythagoras:

$$h^2 + \left(\frac{1}{2}\right)^2 = 1^2 - \left(\frac{1}{2}\right)^2 - \left(\frac{1}{2}\right)^2.$$

So $h = \frac{\sqrt{2}}{2}$ and the sought volume is:

$$\mathscr{V} = \frac{1}{3} \times 1^2 \times \frac{\sqrt{2}}{2} = \frac{2\sqrt{2}}{3}.$$

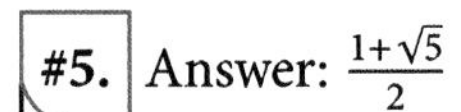

#5. Answer: $\frac{1+\sqrt{5}}{2}$

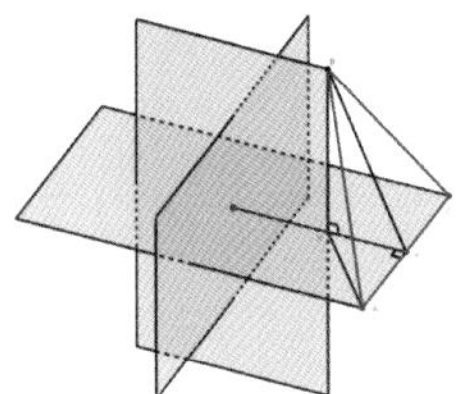

The triangles formed by vertices on different rectangles are always equilateral, let's focus on the isosceles ΔABC.
Say $AC = 2AK = 2$ and $HB = x$. Using twice Pythagoras:

$$AB^2 = AH^2 + HB^2 = AK^2 + KH^2 + HB^2.$$

So $AB^2 = 1^2 + (x-1)^2 + x^2 = 2x^2 - 2x + 2$. We want $AB = AC$ or equivalently $AB^2 = AC^2$, that is $2x^2 - 2x + 2 = 4$ i.e. $x^2 - x - 1 = 0$. This equation has a one positive root known as the golden ratio $x = \frac{1+\sqrt{5}}{2}$. One can check that this value for the sought ratio of sides does indeed give equilateral triangles.

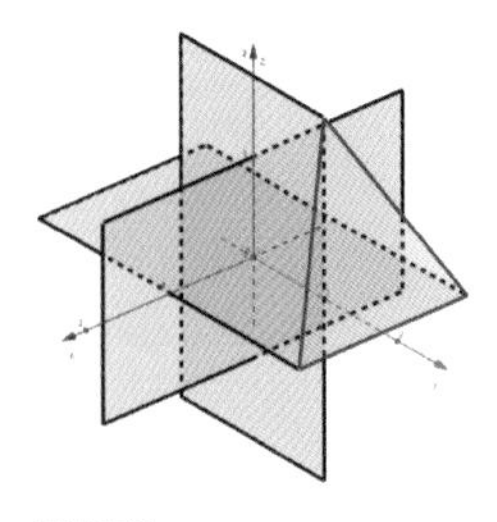

Using the coordinate system shown, the vertices of the blue triangle have coordinates in the form $A(1, x, 0)$, $C(-1, x, 0)$ and $B(0, 1, x)$. The condition $AB = AC$ gives

$$1^2 + (x-1)^2 + x^2 = 2^2 + 0^2 + 0^2$$

i.e. $x^2 - x - 1 = 0$ whose only positive root is the golden ratio.

#6. Answer: $\frac{1}{2}$

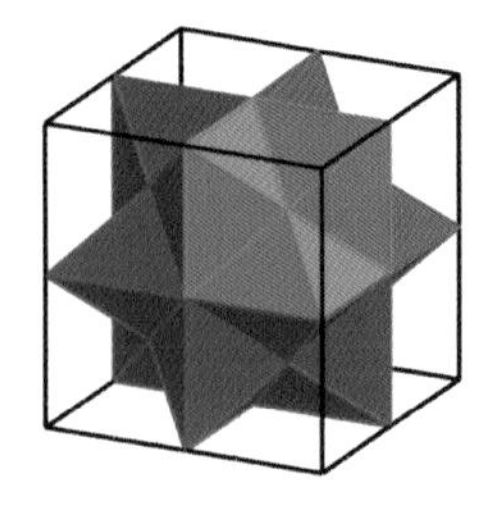

This star is made of 8 identical blocks as the one highlighted. These are in small cubes obtained by splitting the cube in half three times. These blocks can be seen as three square based pyramids with the same apex at the centre of the small cube. The cube is filled by 6 of these pyramids so the volume of the star is one half of the cube's volume.

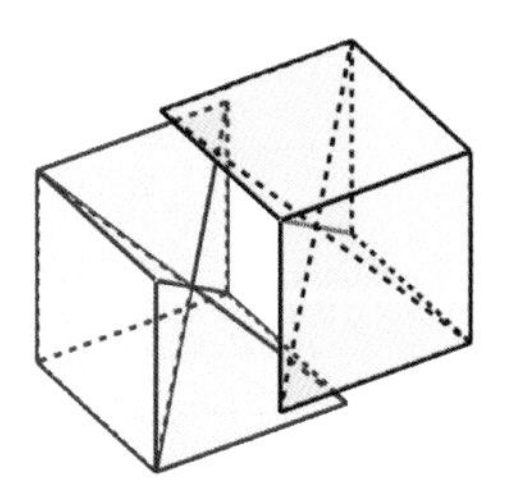

Each 8th of this star gives a block shown previously. Two of these can interlock to complete a cube as shown. Hence the volume of the star is half that of the cube.

#7. Answer: B_2

To find the shortest distance by staying on the surface of the cube, consider the two possible nets of the corner of the cube and superimpose them as shown.

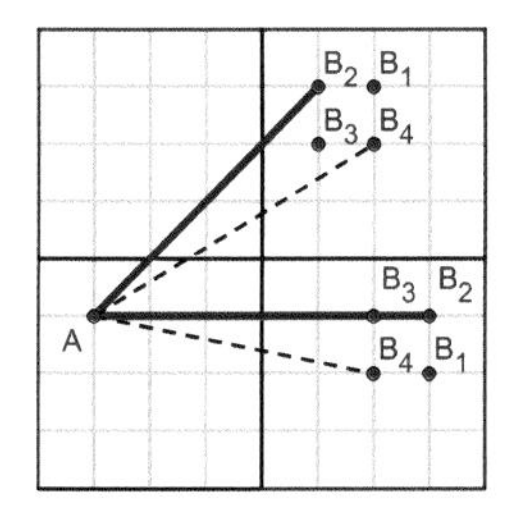

Two examples are shown here:

- The length to connect A to B_2 gives either 6 or $\sqrt{4^2 + 4^2} = \sqrt{32}$. So going through the top face is shorter in this case.
- To connect A to B_4 you get either $\sqrt{5^2 + 1^2} = \sqrt{26}$ or $\sqrt{5^2 + 3^2} = \sqrt{34}$. So going through the top face is longer in this case, as it is for B_1. For B_3, both strategies gives a length of 5.

In the table below, the two measures are compared: the first one being without using the top face.

	B_1	B_2	B_3	B_4
A	$\sqrt{37} < \sqrt{41}$	$\mathbf{6 > \sqrt{32}}$	$5 = 5$	$\sqrt{26} < \sqrt{34}$

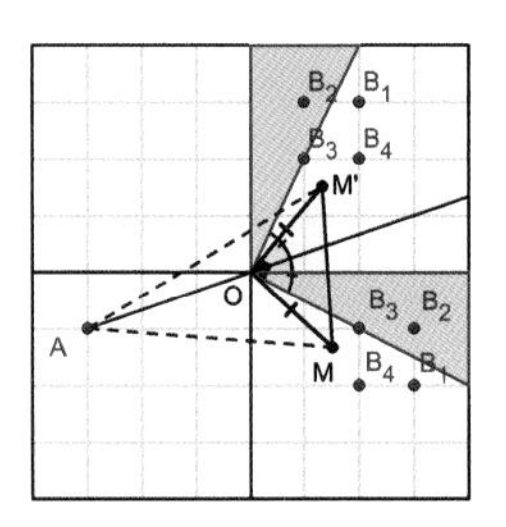

One might want to be more general and find the set of points on the right face for which it is shorter to use the top face when connecting them to A. This zone is shaded here. For a point M in the bottom right square, rotate it 90° about O to get M'. $AM = AM'$ *iff* ΔAOM and $\Delta AOM'$ are congruent, i.e. (AO) is the bisector of $\angle MOM'$. The line (AO) makes a 45° angle with (OB_3) as seen in snack #3 (chap. 2) which gives the limit of the sought zone.

#8. Answer: No, it's impossible.

As we saw in snack #5, the vertices of an icosahedron are the vertices of three perpendicular golden rectangles with a common centre. Let's check if $OABC$ is a regular tetrahedron.

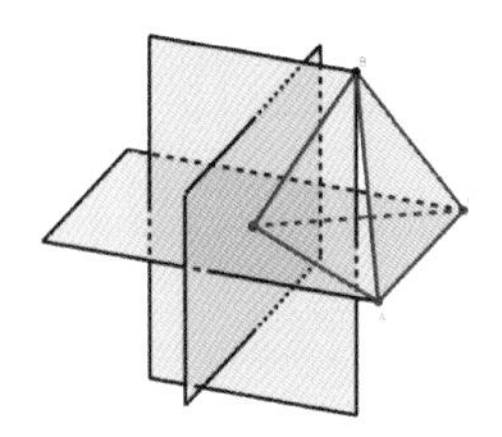

• Let's assume that $AC = AB = BC = 2$, by the pythagorean theorem we have $OA^2 = 1^2 + \varphi^2 = 1 + \varphi + 1 = 2 + \varphi \approx 3.618$. So $OA \neq 2$ and $OABC$ is not a regular tetrahedron.

• Another method is to check if $\angle AOC$ measures 60°. The cosine rule in ΔAOC gives:

$$AC^2 = 2OA^2(1 - \cos(\angle AOC)).$$

So $\cos(\angle AOC) = 1 - \frac{AC^2}{2OA^2} = 1 - \frac{4}{4+2\varphi} = \frac{\varphi}{2+\varphi}$. Or using the dot product:

$$\cos(\angle AOC) = \frac{\overrightarrow{OA} \cdot \overrightarrow{OC}}{OA \times OC} = \frac{\varphi^2 - 1}{2 + \varphi} = \frac{\varphi}{2 + \varphi}.$$

$\varphi \neq 2$, so $\cos(\angle AOC) \neq 1/2$ and $\angle AOC \neq 60°$. Actually, $\angle AOC \approx 63°$.

#9. Answer: Yes, it's possible!

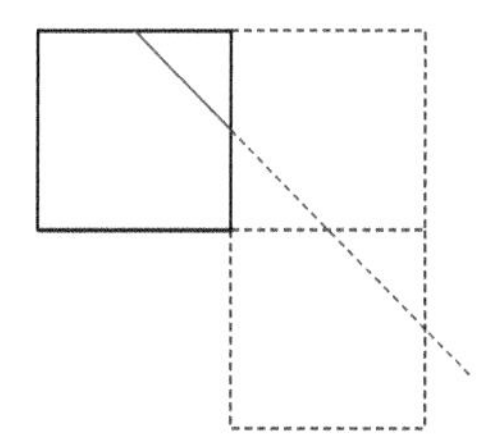

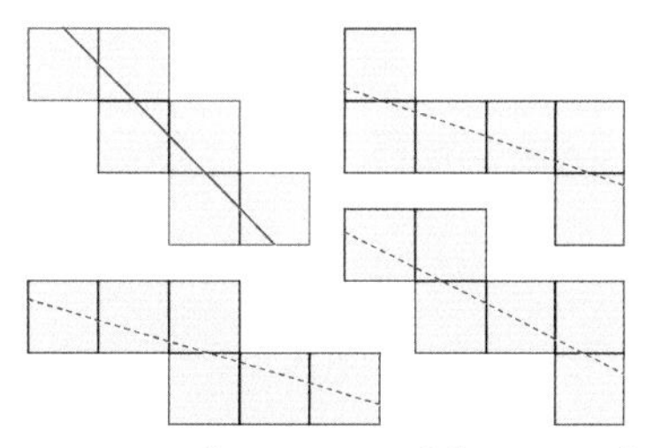

Notice that all the faces are identical: a square with a segment joining the midpoints of two adjacent sides. If you try to place these six pieces to make a straight line, there is just one pattern possible. This pattern (shown here) does correspond to a net of the cube.

Among the 11 possible nets of a cube, only the four shown here allow a straight line to go through all six faces. And among these four nets, the first one is the only one for which the line intersects all the edges in their midpoint. And if you try to fold them, you'll see it's also the only one for which it makes a closed figure.

#10. Answer: B

Consider the central triangle as the base of the tetrahedron. The apex is on the intersection of three planes perpendicular to the dashed sides going through the vertices of the big triangle. From above you see the heights of ΔABC. The foot of the height of the tetrahedron is therefore the orthocentre H of ΔABC. Since the base has constant area, we want the greatest height. Consider the top flap EIJ, its altitude from E is always 4. When folding it, the height of the tetrahedron can never be greater than 4, which occurs when this face is perpendicular to the base, i.e. when H is on (IJ).

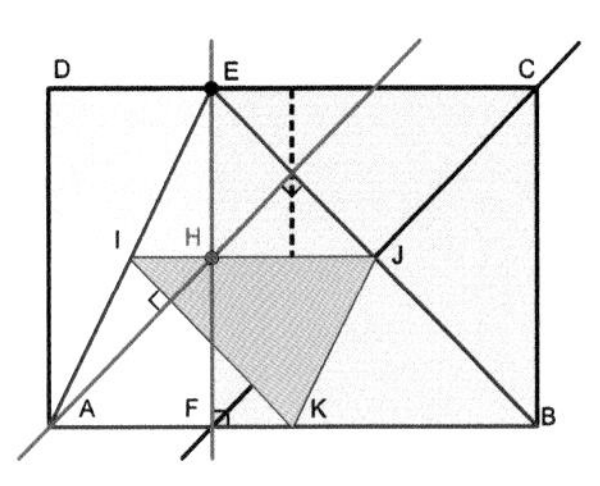

This happens for the second figure when E is at one third of the long side of the rectangle. Indeed, by splitting the rectangle into three squares (shown shaded) you'll see two heights meet on (IJ) in that case.

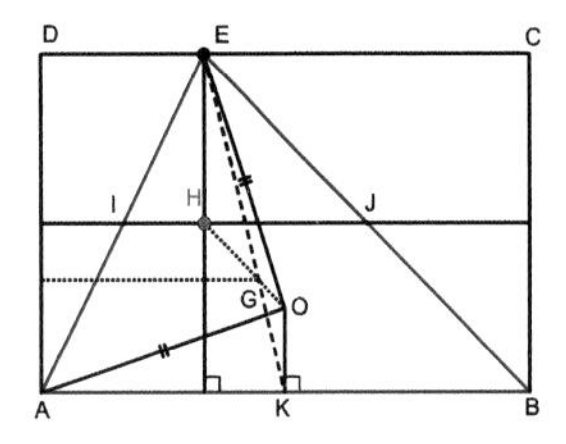

Let's use a natural coordinate system where A is the origin. The centroid G of ΔABE is such that $y_G = 8/3$. The circumcentre O being on the perpendicular bisector of AB, $x_O = 6$. We know that O, G and H are collinear with $GH = 2.OG$. The constraint of H being on (IJ) means $y_H = 4$. From this we deduce that $y_O = 2$, so $O(6, 2)$. $OE^2 = OA^2$ for $E(x, 8)$ gives:

$$(6 - x)^2 + (8 - 2)^2 = 6^2 + 2^2 \iff (6 - x)^2 = 4 \iff x = 6 \pm 2.$$

So the sought position for E is at one third of CD.
For a general construction, place O at the quarter of the width of the rectangle and draw the circle with centre O going through A. If it intercepts the top side in two points E and E', then the net ABE is the net of a tetrahedron with the greatest volume.

Curiosity Closet

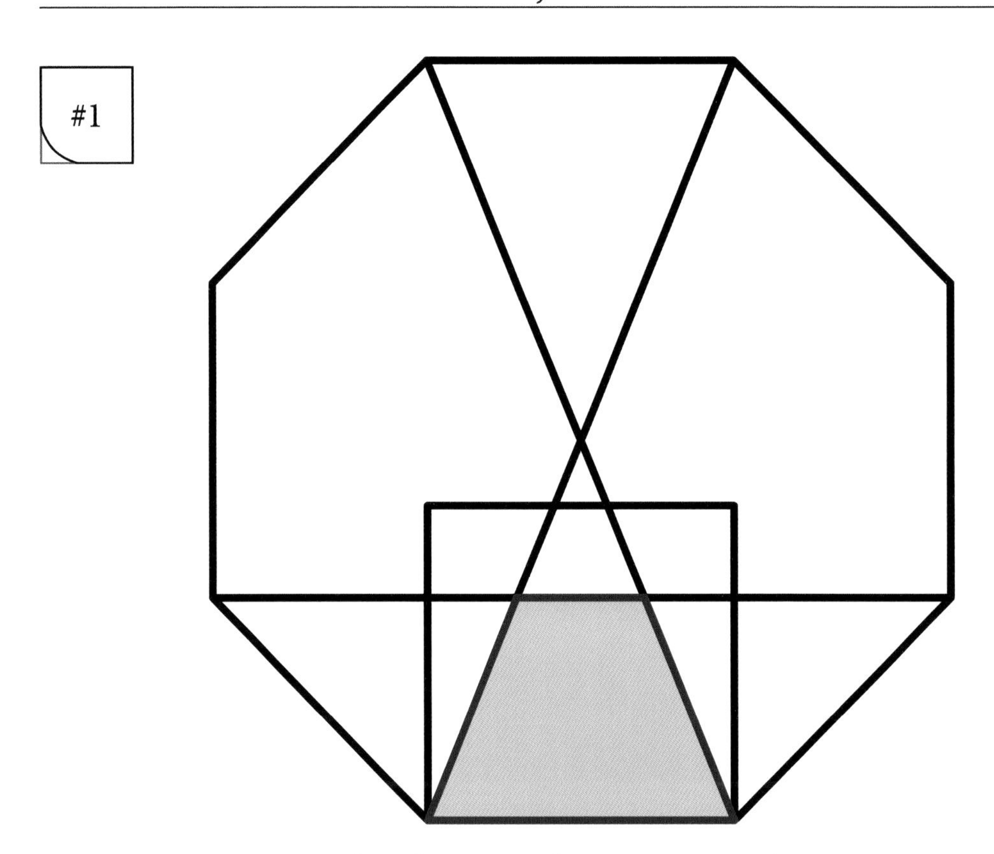

A square is constructed within a regular octagon as shown.

What fraction of the square is shaded?

#2

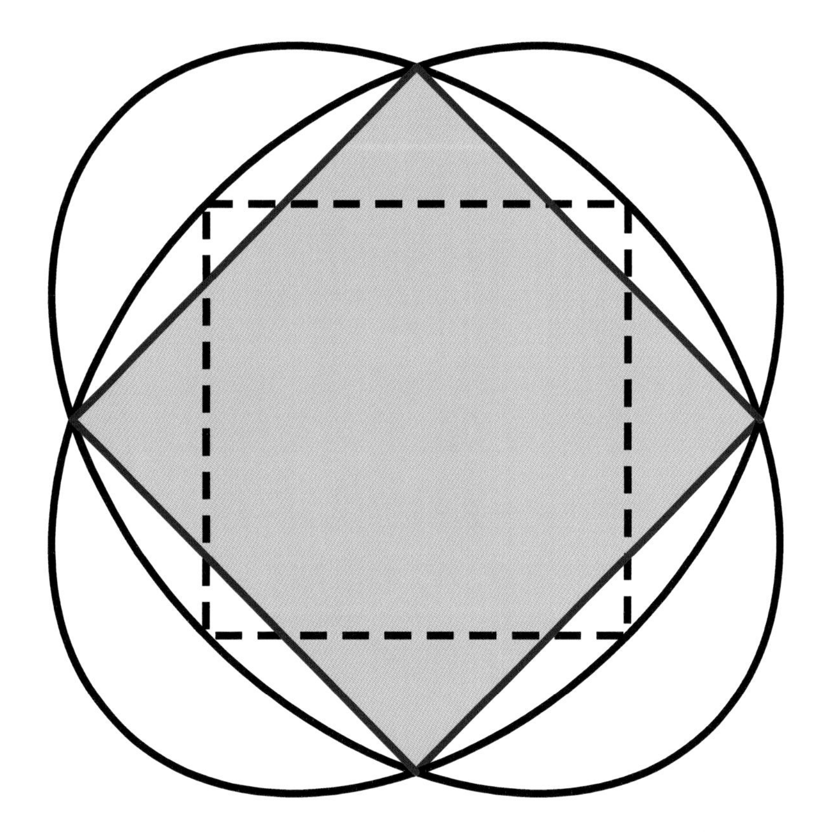

Two ellipses are constructed using opposite vertices of the dotted square as foci such that the remaining vertices also lie on each ellipse. A second square is then constructed using the points where the two ellipses intersect. What is the ratio of the areas of each square?

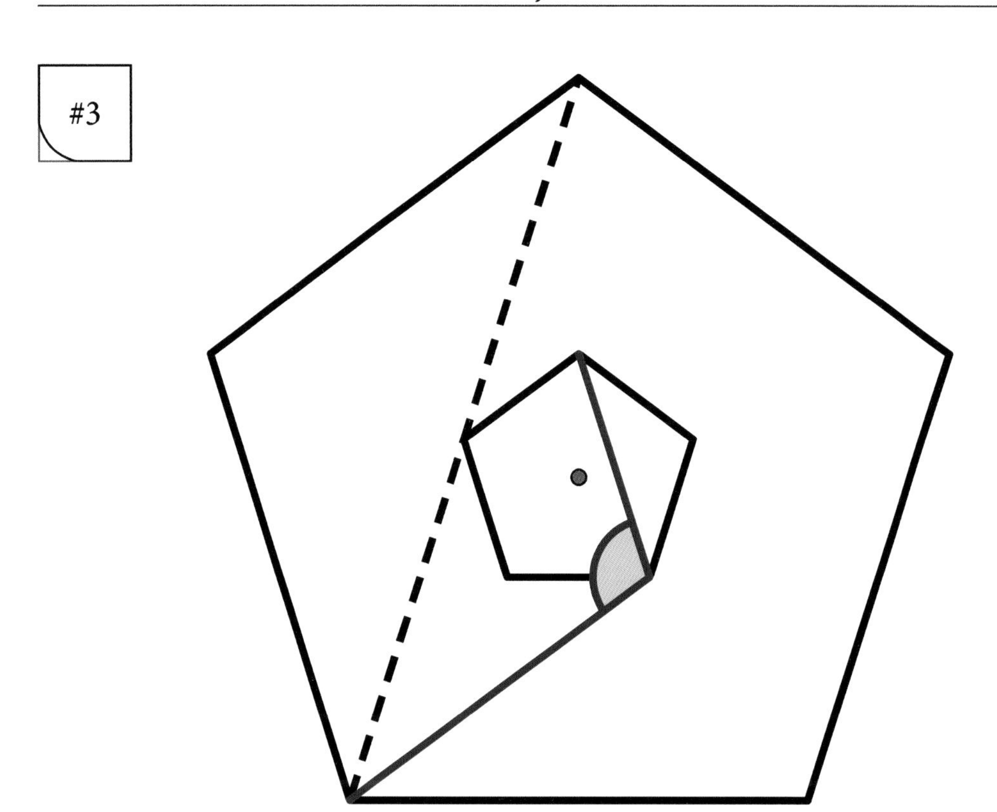

Two concentric regular pentagons are constructed with parallel sides such that the upper left vertex of the smaller pentagon is on a diagonal of the big one.

What is the missing angle?

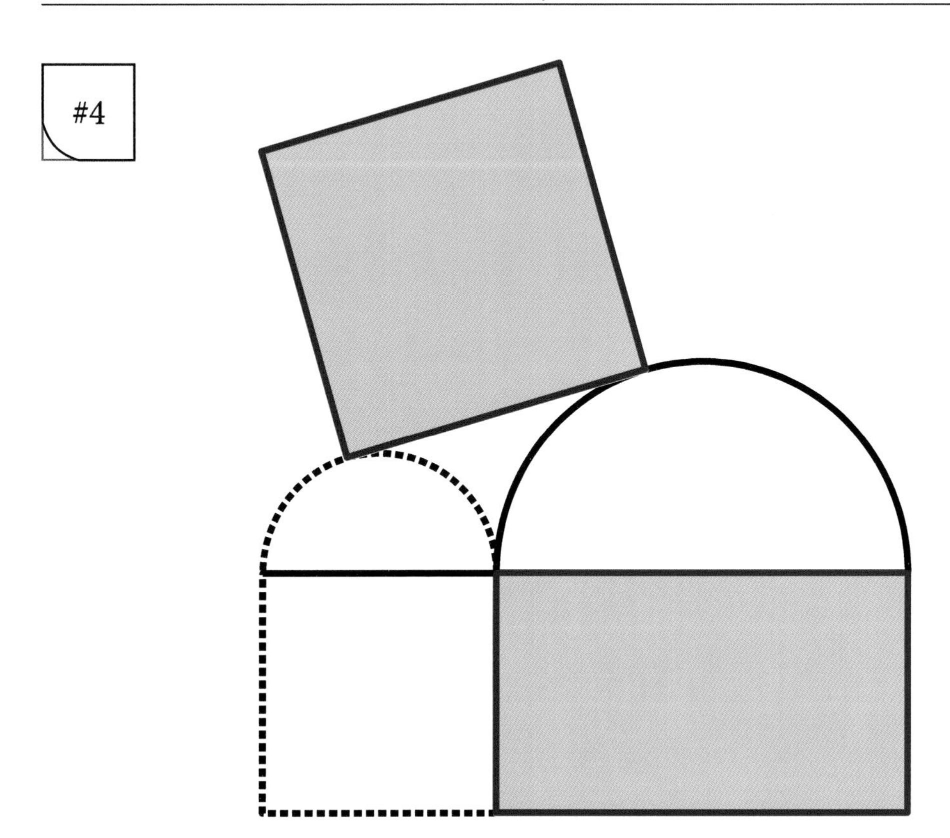

Two semicircles lie on a straight line as shown. A square is constructed on the tangent line between them. A dotted square and shaded rectangle are constructed using the diameters of the semicircles.

Show that the two shaded areas are equal.

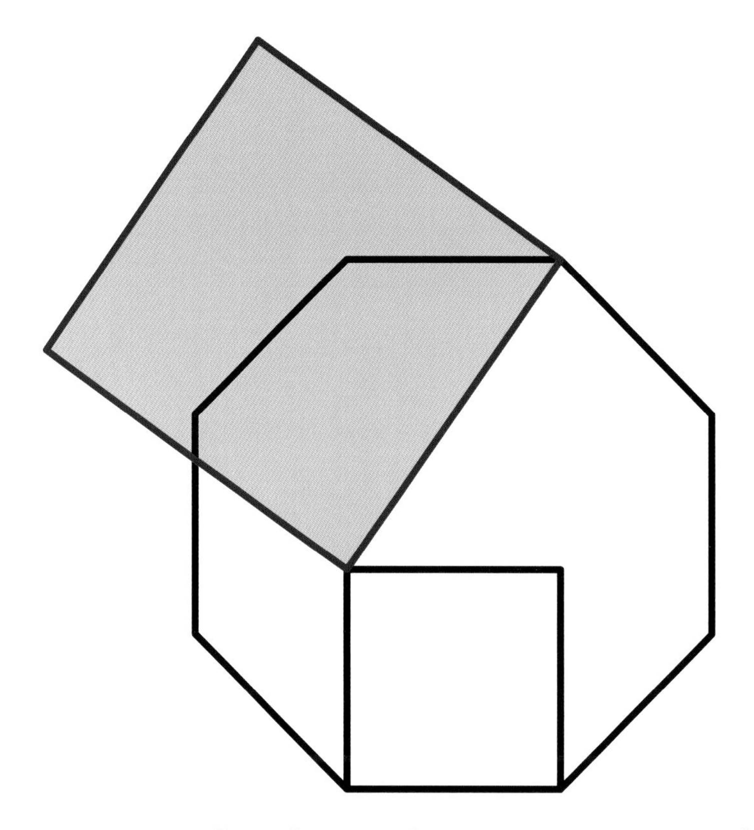

A unit square is constructed inside a regular octagon using one of its sides. A second shaded square is constructed using a vertex of the original square and a vertex of the octagon as shown.

What is the ratio of the areas of the two squares?

#6

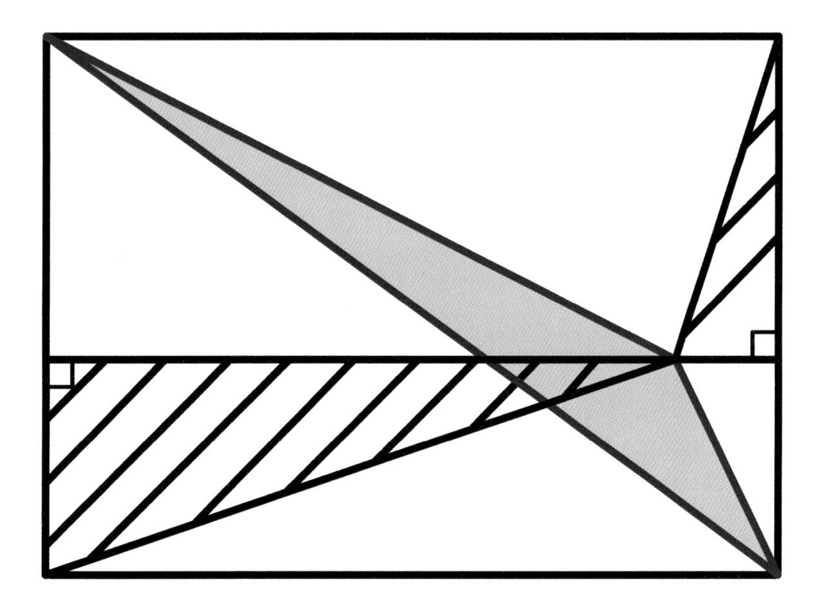

From a point inside a rectangle, a perpendicular to the sides is drawn.

Prove that the area of the shaded triangle is equal to the difference of the areas of the two hatched triangles.

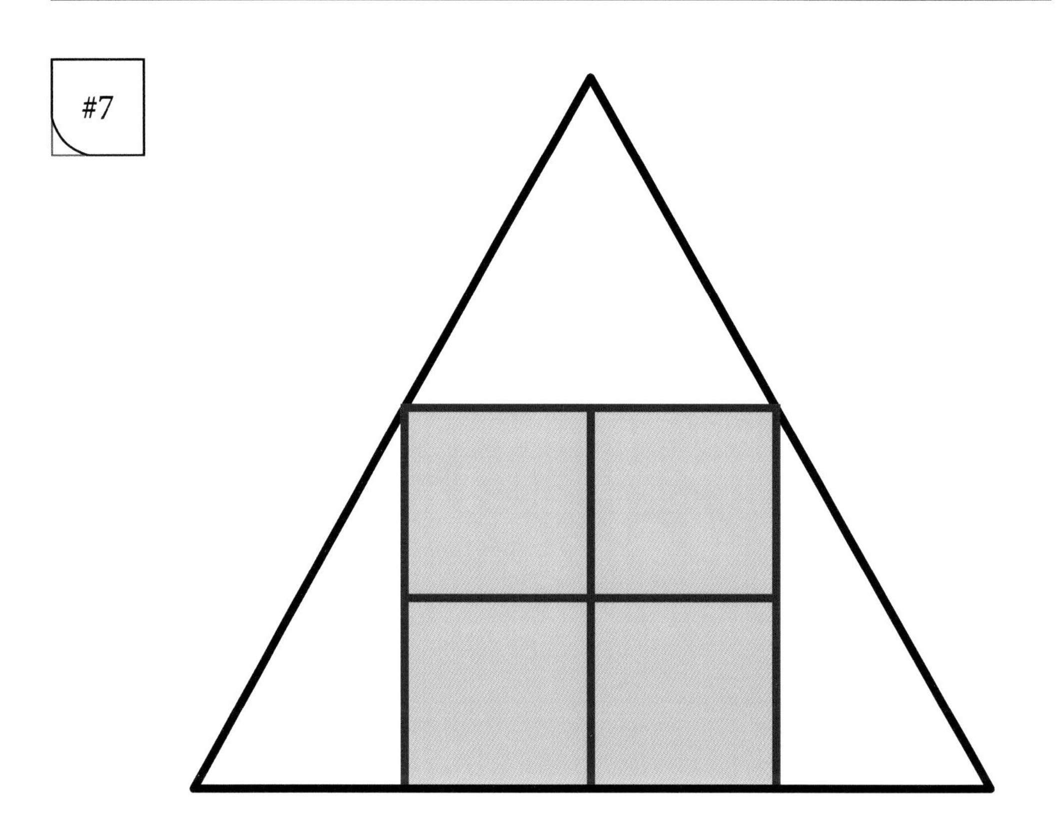

Four unit squares have been stacked so that the two top vertices are on the sides of the equilateral triangle.

Can you fit more unit squares in this equilateral triangle?

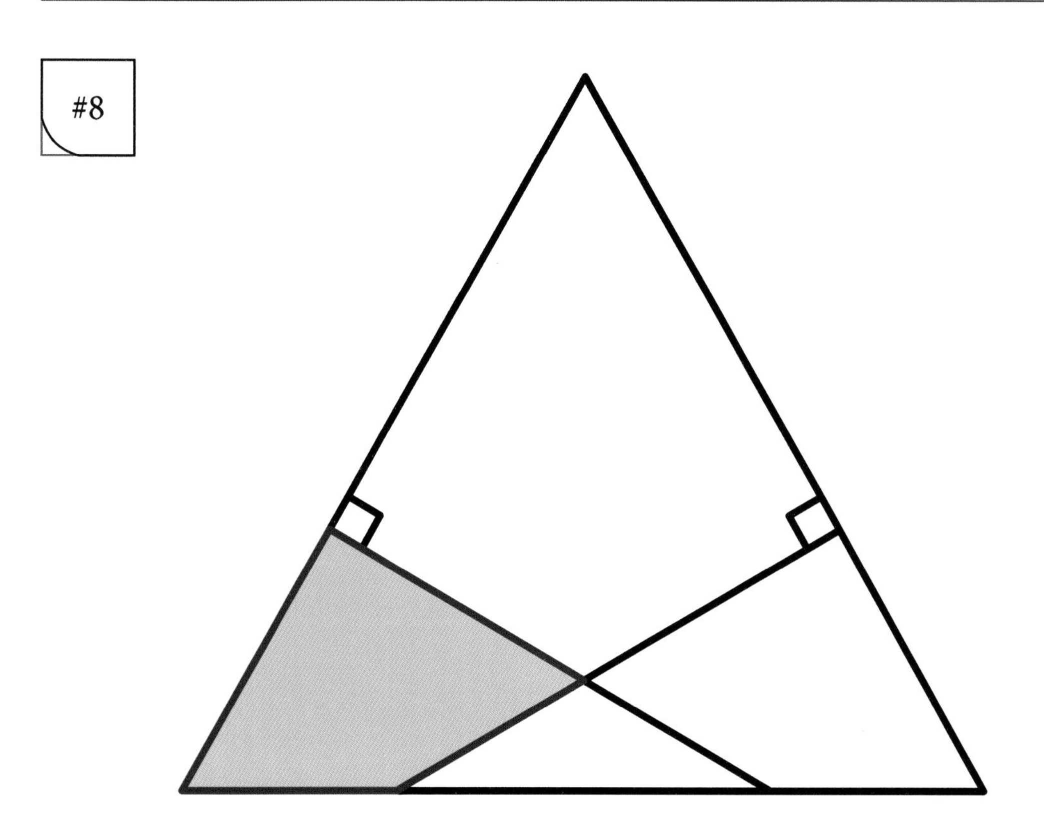

In an equilateral triangle, two lines perpendicular to the sides form two congruent kites.

What fraction of the equilateral triangle is shaded?

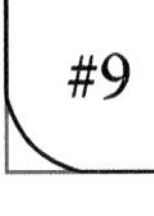

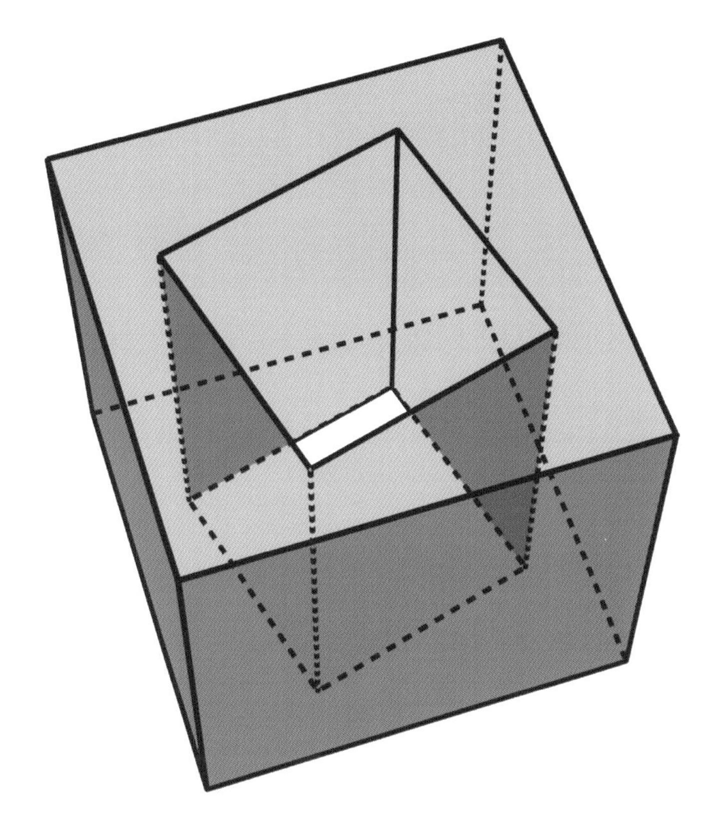

A square hole has been drilled through a cube, allowing a smaller cube to go through.

How big is the largest cube you can push through by drilling a square hole through the centre of the cube, without breaking it apart?

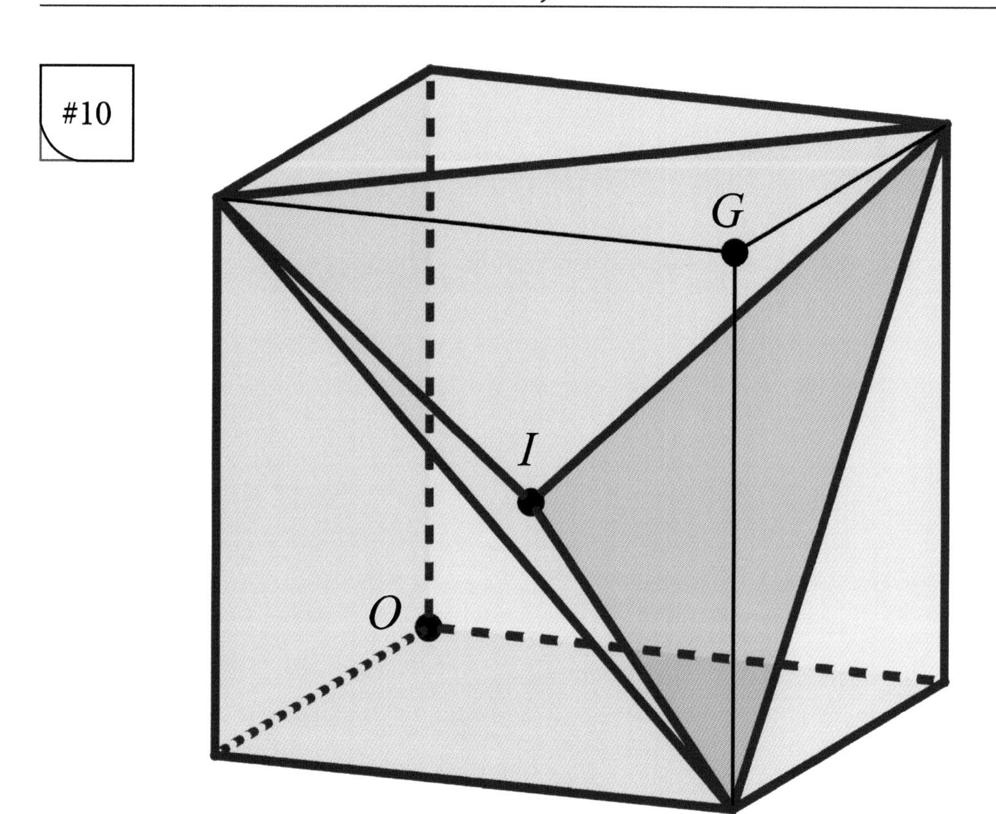

A vertex I of a unit cube has been punched in so that it is still at a unit distance from its three neighbours but inside the cube.

What is the ratio $OI : OG$?

Solutions and Answers

#1. Answer: $\frac{1}{2}$

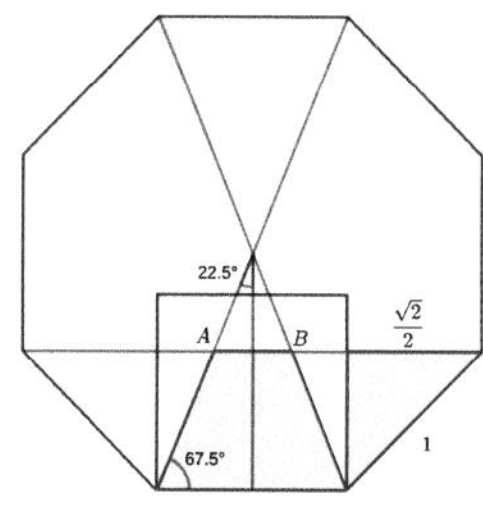

Using the triangle to the right of the square, we can determine the height of the trapezium using Pythagoras as $\frac{\sqrt{2}}{2}$. We can then calculate the length AB by finding the full height of the triangle that is formed by the base and centre of the octagon: then use the properties of similar triangles to subtract the area of the smaller similar triangle from the greater one.

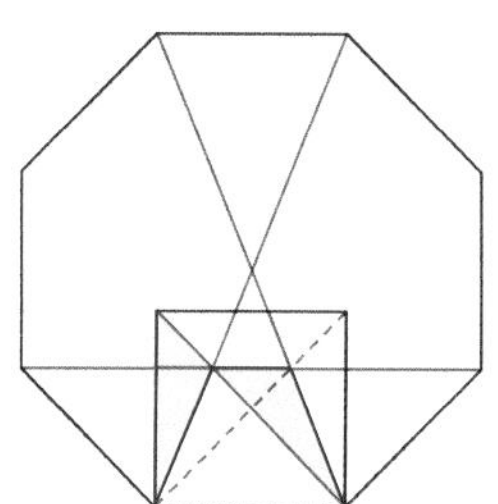

By drawing in the diagonals of the square, it becomes apparent that the shaded triangles indicated are congruent, hence the original shaded area is equal to that of the triangle made by a diagonal across the square.

#2. Answer: 3 : 4

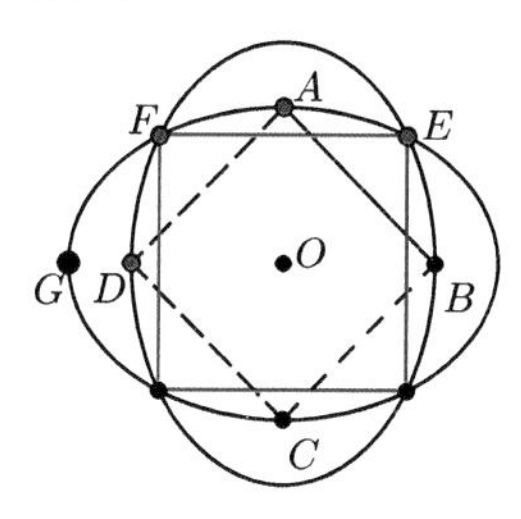

Reconstruct the diagram on a coordinate grid as shown. The centre of the square is $O(0,0)$ and $AB = 2$. Using $OA^2 = OG^2 - OB^2$ we can determine that $OG = 4$. So an equation of the ellipse is $\frac{x^2}{4} + \frac{y^2}{2} = 1$ which simplifies to $x^2 + 2y^2 = 4$. The two ellipses intercept in E with $y = x$ so $3x^2 = 4$. So $x_E = \frac{2}{\sqrt{3}}$ and the area of the second square is $\left(\frac{4}{\sqrt{3}}\right)^2$ and so the ratio between the two squares is 3 : 4.

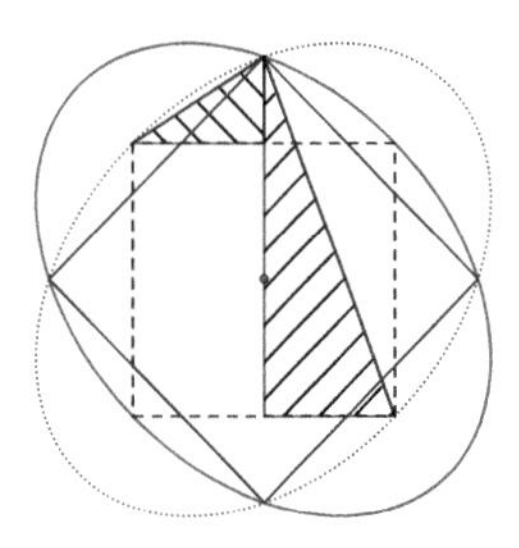

The sum of the distances from a point on an ellipse to the two focal points is constant. Say the dashed square has a side length of 2, we want the sum of the two hypotenuses of the hatched triangles to be equal to 2+2 = 4. Say the shortest leg measures x, we have:

$$\begin{aligned}\sqrt{x^2+1}+\sqrt{(x+2)^2+1}=4 &\iff (x+2)^2+1=4^2+x^2+1-2\times 4\times\sqrt{x^2+1}\\ &\iff 4x-12=-8\sqrt{x^2+1}\\ &\iff (4x-12)^2=64(x^2+1)\\ &\iff 48x^2+96x-80=0\\ &\iff x^2+2x-\frac{5}{3}=0\\ &\iff (x+1)^2=\frac{8}{3}\end{aligned}$$

No need to solve this equation because $(x+1)$ is half the diagonal of the shaded square, so its area is $2(x+1)^2=\frac{16}{3}$. And since the dashed square has an area of $2^2=4$, the sought ratio of areas is 3 : 4.

#3. Answer: 108°

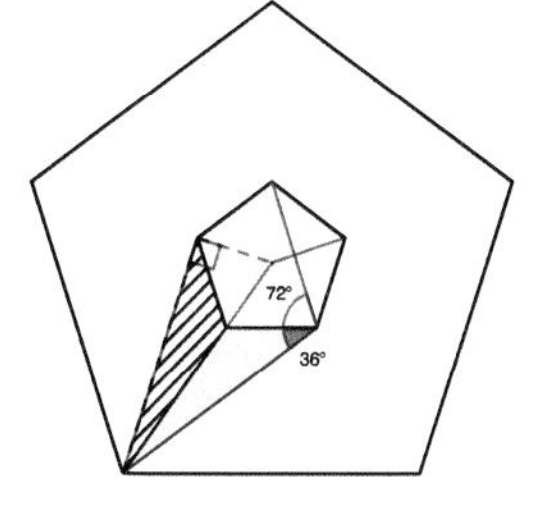

The additional construction lines shown highlight two shaded congruent triangles. We can determine the angle shown of 36° as it is 90 – 54. From here we can calculate the missing angle as 108°.

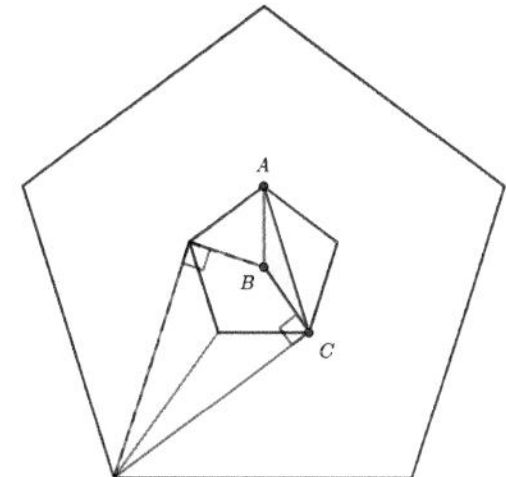

Proof without words

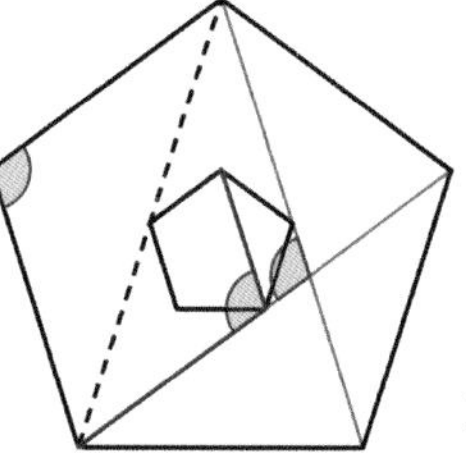

Proof without words

Curious: This is the same as the internal angle of a regular pentagon.

#4. Answer: They are equal.

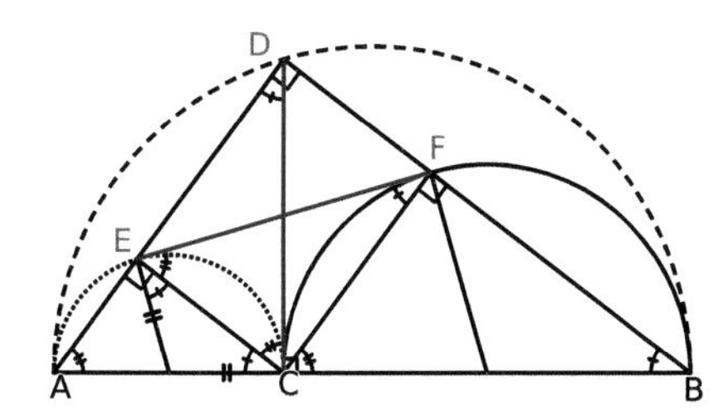

Draw the perpendicular to *AB* through C which intersects the dashed semicircle of diameter *AB* in *D*. Draw the lines *DA* and *DB* which intersect the semicircles in *E* and *F*. *DECF* has three right angles so it is a rectangle, and *EF* is the tangent to both semicircles (proof left to the reader using the marked angles).

We are asked to prove that $AC \times CB = EF^2$. Since ACD and DCB are similar, $\frac{CD}{AC} = \frac{CB}{CD}$ so $CD^2 = AC \times CB$. Finally, $CD^2 = EF^2$ because EF and CD are the diagonals of the rectangle $DECF$.

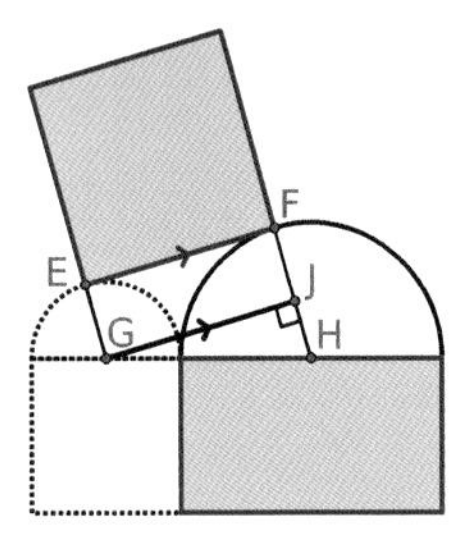

Let the radii of the small and large semicircles be r_1 and r_2 respectively. Construct a parallel line to EF from G and extend two sides of the shaded square down to the centre points of the two circles. Using newly constructed triangle HGJ we can use Pythagoras to determine that the area of the square is

$$GJ^2 = (r_2 + r_1)^2 - (r_2 - r_1)^2 = 4r_1r_2.$$

We can also see that the area of the blue rectangle is $2r_2 \times 2r_1 = 4r_1r_2$.

Curious: This shows that this is a construction of a square whose side length is the *geometric mean* of the two diameters (*i.e.* of the length and width of the shaded rectangle).

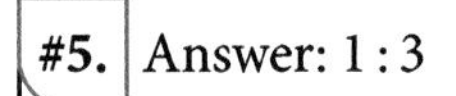

Answer: 1 : 3

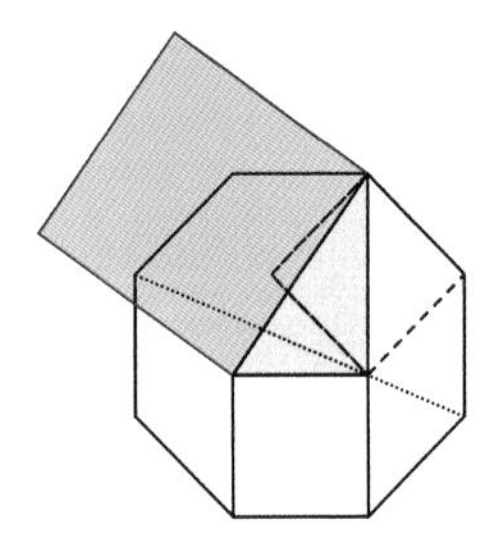

Construct a right triangle as shown, with base 1, its height is $\sqrt{2}$ as is shown using the dotted mirror line and the dashed square. Therefore the hypotenuse of the constructed triangle is $\sqrt{3}$ so the area of the larger square is 3.

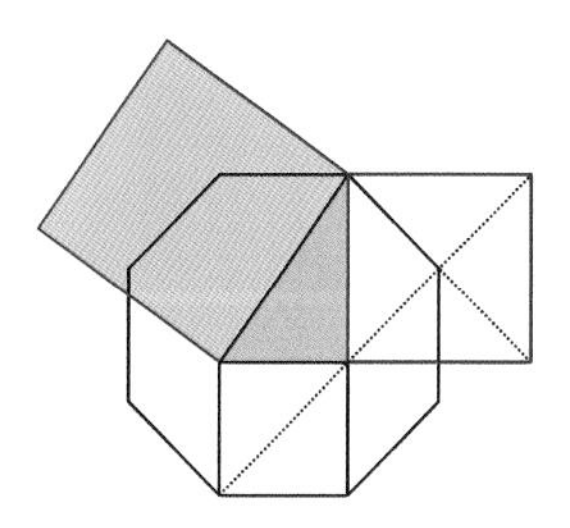

This figure shows that the two white squares constructed on the legs of the shaded right angled triangle have an area of 1 and 2 respectively, so the shaded one has an area of 3 by Pythagoras.

#6. Answer:

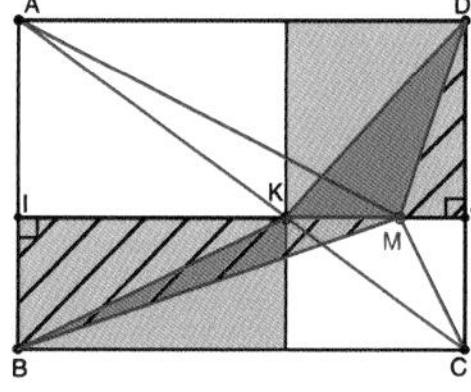

We'll write $[XYZ]$ for the area of a triangle XYZ. Draw a vertical line through K where the horizontal line intersects the diagonal. The diagonal of the rectangle divides it in two congruent triangles, so the two dark rectangles have the same area, so $[DKJ] = [BKI]$. Notice that $[AKM] = [DKM]$ and $[DKM] + [DMJ] = [DKJ]$. Now the bottom part: $[BKI] = [BMI] - [BMK]$. $[DKJ] = [BKI]$, so we have $[DKM] + [DMJ] = [BMI] - [BMK]$ i.e. $[DKM] + [BMK] = [BMI] - [DMJ]$ or: $[AMC] = [BMI] - [DMJ]$.

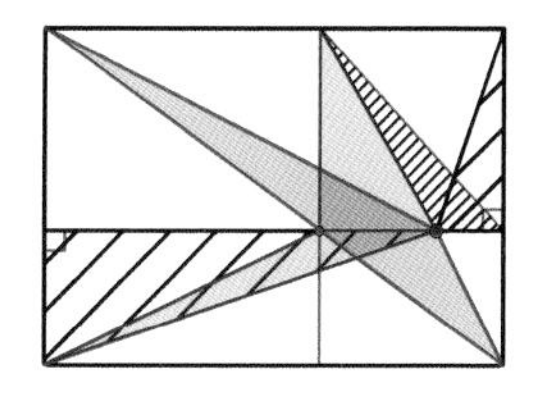

We used three shearings: on the top the hatched triangle and the shaded have been sheared to have a common vertex showing that their sum is half of the top right rectangle. That is equal to half the bottom left rectangle which appears to be the difference of the hatched and the shaded triangle. So the difference of the areas of the two hatched triangles equals the area of the shaded triangle.

Curious: This gives a necessary and sufficient condition using areas for a point to be on the diagonal of a rectangle.

#7. Answer: 6

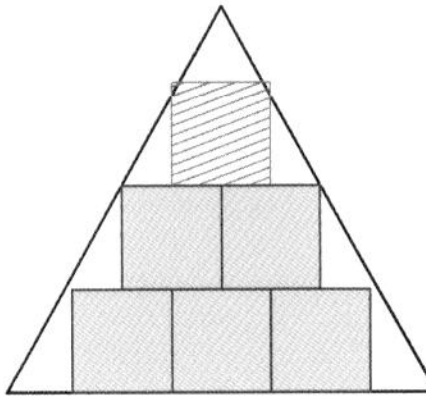

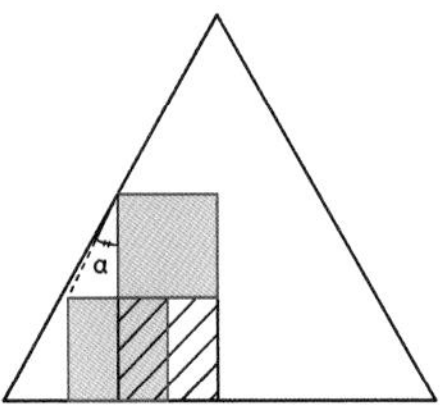

The first figure shows that you can place three unit squares on the base of the equilateral triangle but not another on top of this stack. To check that this is possible, consider the angle α in the second figure, its tangent is $\frac{1}{2}$ which is smaller than $\tan(30°) = \frac{1/2}{\sqrt{3}/2}$. So $\alpha < 30°$ and you can slide the bottom hatched square half a unit to the left and similarily for the one on the right, leaving space for a third one in the middle.

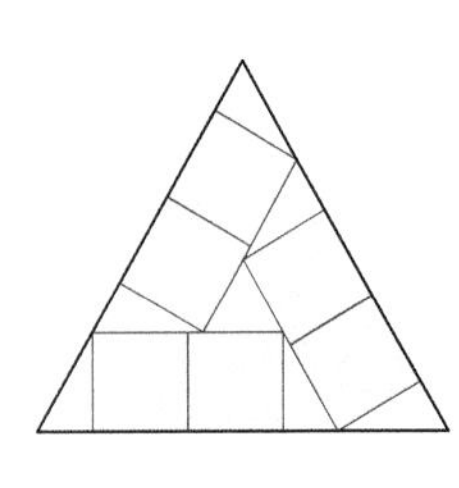

This figure shows a way to place 6 unit squares in the equilateral triangle. Indeed, from the initial figure you can determine that the side of the triangle is $2 + 2 \times 2\tan(30°) = 2 + \frac{4}{\sqrt{3}}$.

Since $\frac{1}{\sin(30°)} = \frac{2}{\sqrt{3}} = 2\tan(30°)$, this can be written as $2 + 2 \times \tan(30°) + \frac{1}{\sin(30°)}$ which corresponds to the length on the side shown on the figure with six unit squares.

Curious: It's surprising that the same triangle gives the highest packing density for 5 and 6 unit squares inside an equilateral triangle.[1]

[1] Found by Erich Friedman in 1997. See https://www2.stetson.edu/~efriedma/squintri/.

#8. Answer: $3\sqrt{3} - 5$

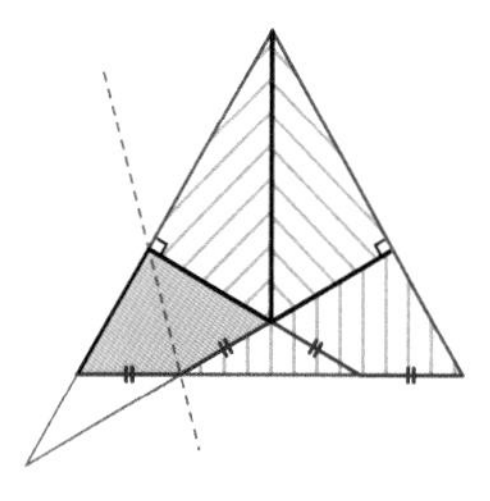

By calculating angles, and using lines of symmetry (the dashed line for the shaded region and the vertical mirror line of the equilateral triangle) you'll see you that the shaded region is what is left from the equilateral triangle once you have removed the three hatched congruent 30°, 60°, 90° triangles. Let's consider these right triangles have sides 1, $\sqrt{3}$, 2, their area is then $\mathscr{A}_t = \frac{\sqrt{3}}{2}$ and the equilateral triangle's area is $\mathscr{A}_T = \frac{\sqrt{3}}{4} \times (1 + \sqrt{3})^2$. The shaded fraction is:

$$1 - \frac{3\mathscr{A}_t}{\mathscr{A}_T} = 1 - 3\frac{2}{\left(1 + \sqrt{3}\right)^2} = 1 - 3 \times \left(2 - \sqrt{3}\right) = 3\sqrt{3} - 5.$$

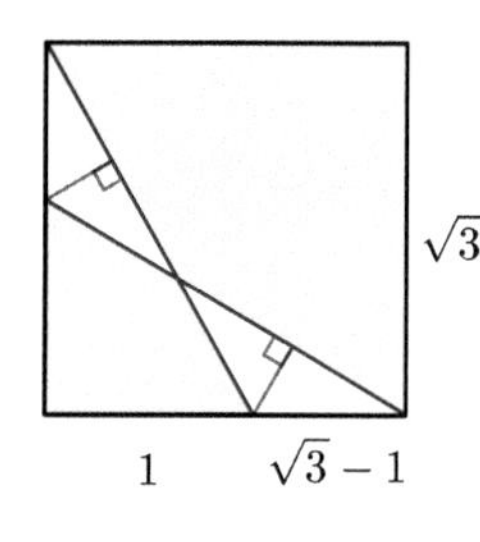

Using the same measures as previously, the shaded kite can be seen as two overlapping congruent 30°, 60°, 90° triangles and the figure can be prolonged into a square as shown. Let x be the area of the smaller kite, a similar kite of area $3x$ is formed. The remaining white zone in the square represents the area of two equilateral triangles of side $\sqrt{3} - 1$ so we have the equation:

$$x + 3x + 2 \times \frac{\sqrt{3}}{4} \times \left(\sqrt{3} - 1\right)^2 = \sqrt{3}^2$$

We find $x = \frac{3+\sqrt{3}}{8}$ so the sought fraction is $x \div \mathscr{A}_T = x \div \left(\frac{\sqrt{3}}{4} \times \left(1 + \sqrt{3}\right)^2\right) = 3\sqrt{3} - 5$.

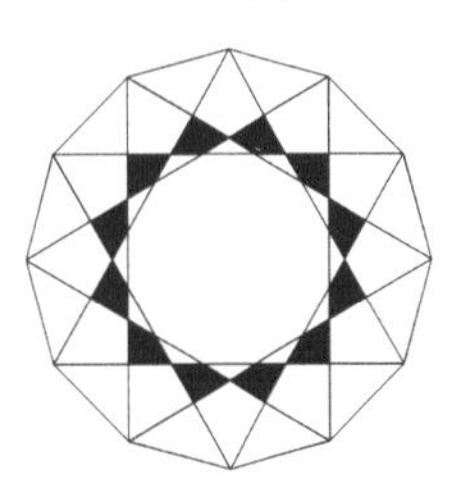

Curious: This kite can appear when permorming four 30° rotations of an equilateral triangle (or three 30° rotations of a square) about its centre, forming a regular dodecagon.

#9. Answer: A cube larger than the original one can go through.

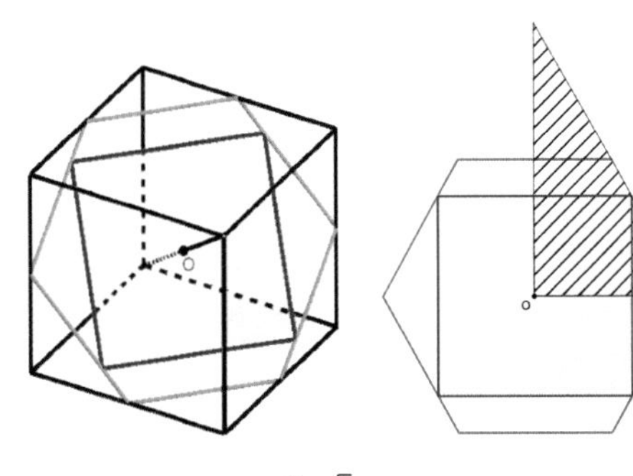

The first idea one might have is to drill a hole using the direction of the great diagonal. We are searching for the largest square we can fit in a regular hexagon. Let x be half the side of the square and suppose that the hexagon has a unit side length. Using the similar 30°, 60°, 90° right triangles shaded and hatched, we can write the ratio of the legs: $1 - x = \frac{x}{\sqrt{3}}$. Solving for x gives $x = \frac{3-\sqrt{3}}{2}$ so the square has an area of $(2x)^2 = (3 - \sqrt{3})^2 = 12 - 6\sqrt{3}$. If the cube has unit side length, the hexagon has sides measuring $\frac{1}{\sqrt{2}}$ so its area is half what we considered and the square has an area of $6 - 3\sqrt{3}$. So the edge is $\sqrt{6 - 3\sqrt{3}} \approx 0.9$. That's almost 90% of the original cube's edge length.

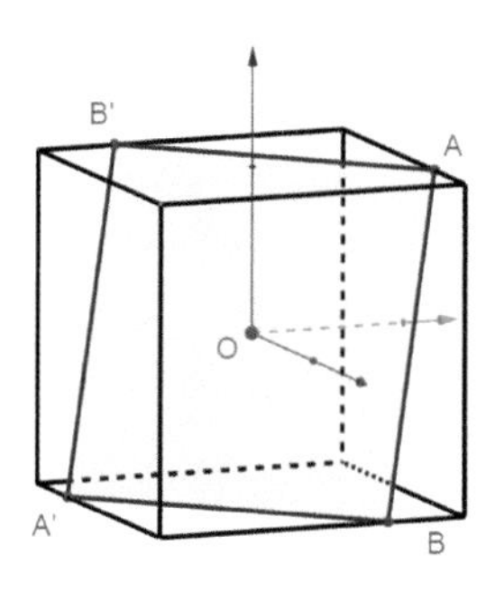

Let's consider the cube centered at the origin of an orthonormal coordinate system with vertices $(\pm 1, \pm 1, \pm 1)$. Starting from a point $A(a, 1, 1)$ on a edge (with $0 < a < 1$), we consider $A'(-a, -1, -1)$, $B(1, a, -1)$ and $B'(-1, -a, 1)$. Then $ABA'B'$ is a rectangle (parallel sides and congruent diagonals) and it is a square *iff* $AB = AB'$ i.e.

$$2(a-1)^2 + 2^2 = 2(a+1)^2 \iff (a+1)^2 - (a-1)^2 = 2 \iff 4a = 2 \iff a = \frac{1}{2}.$$

The area of our square is $AB'^2 = 2(a+1)^2 = \frac{9}{2} = 4.5$. Since we started with a cube of edge 2, the area of a face is 4. So the ratio of the edges is $\sqrt{\frac{9/2}{4}} = \frac{3\sqrt{2}}{4} \approx 1.06$

Curious: We managed to make a cube 6% *larger* than the original one go through it! This is known as *Prince Rupert's cube*.

#10. Answer: 1 : 3

Label A, B, C the three neighbours of I and G. Since they all satisfy $MI = MG$, (IG) is perpendicular to (ABC) and (ABC) intersects $[IG]$ in its midpoint. I is therefore the reflection of G about this plane (ABC).

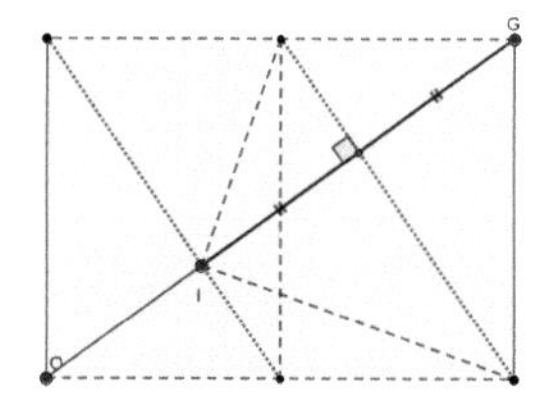

• This is the view you have in the plane splitting the cube in half along a diagonal containing O, G and I. Because of the reflection mentioned previously, you have the equal lengths marked. And because the figure is symmetrical about the centre of the rectangle, these lengths equal OI. Hence $OI : OG$ is 1 : 3.

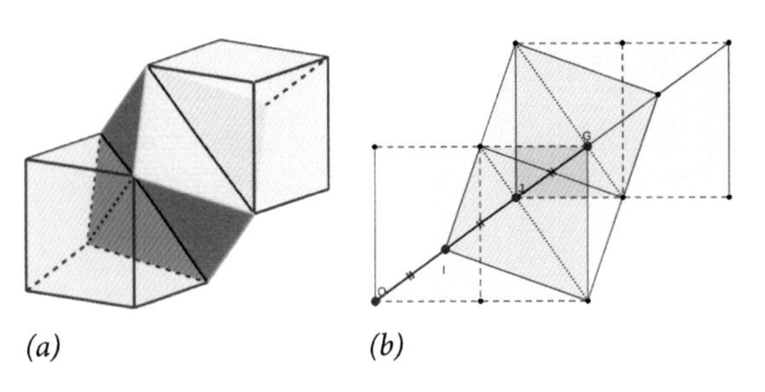

(a) (b)

•• The punched in corner has by symmetry the shape of a negative cube corner. You can fit a grey cube in it and repeat the process, creating a chain of cubes as shown in figure (a). The third cube is a translated version of the original cube along the great diagonal.

Figure (b) shows the same view as previously but the three cubes allow to see why I produces a trisection of the great diagonal.

••• The unit cube offers an orthonormal coordinate system where O is the origin. By a symmetry argument we know that I is on the diagonal (OG). Say $I(x, x, x)$, writing that the distance from I to $(1, 1, 0)$ is 1 gives:

$$2(x-1)^2 + x^2 = 1 \iff 3x^2 - 4x - 1 = 0.$$

This equation has two real roots: 1 and $\frac{1}{3}$. So $I\left(\frac{1}{3}, \frac{1}{3}, \frac{1}{3}\right)$ and $OI : OG$ is $1 : 3$.

Curious: Although we started with a 3D problem, if you look at the first figure, it gives a method to fold an A4 sheet of paper in thirds; in both length and width since the point I is on the diagonal.